T0142098

Advances in Intelligent Systems and Computing

Volume 374

Series editor

Janusz Kacprzyk, Polish Academy of Sciences, Warsaw, Poland
e-mail: kacprzyk@ibspan.waw.pl

About this Series

The series "Advances in Intelligent Systems and Computing" contains publications on theory, applications, and design methods of Intelligent Systems and Intelligent Computing. Virtually all disciplines such as engineering, natural sciences, computer and information science, ICT, economics, business, e-commerce, environment, healthcare, life science are covered. The list of topics spans all the areas of modern intelligent systems and computing.

The publications within "Advances in Intelligent Systems and Computing" are primarily textbooks and proceedings of important conferences, symposia and congresses. They cover significant recent developments in the field, both of a foundational and applicable character. An important characteristic feature of the series is the short publication time and world-wide distribution. This permits a rapid and broad dissemination of research results.

Advisory Board

More information about this series at http://www.springer.com/series/11156

Tania Di Mascio · Rosella Gennari · Pierpaolo Vittorini
Fernando De la Prieta
Editors

Methodologies and Intelligent Systems for Technology Enhanced Learning

 Springer

Editors
Tania Di Mascio
Department of Information Engineering,
    Computer Science and Mathematics
University of L'Aquila
L'Aquila
Italy

Rosella Gennari
Computer Science Faculty
Free University of Bozen-Bolzano
Bolzano
Italy

Pierpaolo Vittorini
Department of Life, Health and
    Environmental Sciences
University of L'Aquila
Coppito
Italy

Fernando De la Prieta
Department of Computer Science
    and Automation Control
University of Salamanca
Salamanca
Spain

ISSN 2194-5357   ISSN 2194-5365   (electronic)
Advances in Intelligent Systems and Computing
ISBN 978-3-319-19631-2   ISBN 978-3-319-19632-9   (eBook)
DOI 10.1007/978-3-319-19632-9

Library of Congress Control Number: 2015940027

Springer Cham Heidelberg New York Dordrecht London

Springer International Publishing AG Switzerland is part of Springer Science+Business Media
(www.springer.com)

Preface

Education is the cornerstone of any society and it is the basis of most of the values and characteristics of that society. Knowledge societies offer significant opportunities for new techniques and applications, especially in the fields of education and learning. Technology Enhanced Learning (TEL) provides knowledge processing technologies to support learning activities in knowledge societies. The role of intelligent systems, rooted in Artificial Intelligence (AI), has become increasingly relevant to shape TEL for knowledge societies. However, AI-based technology alone is clearly not sufficient for improving the learning experience. Design methodologies that are based on evidence and involve end users are essential ingredients as well. Stronger with them, AI-based TEL can cater to its learners' tasks and, at the same time, support learning while performing tasks. In brief, we believe that it is the masterful combination of evidence, users and AI that can allow TEL to provide an enhanced learning experience.

The 2nd MIS4TEL conference will build on the experience matured within the 1st MIS4TEL conference as well as a series of evidence based TEL workshops: ebTEL, ebuTEL & iTEL.

The conference and these proceedings bring together researchers and developers from industry and the academic world to report on the latest AI-based scientific research and advances as well as empirical methodologies for TEL.

This volume presents all papers that were accepted for MIS4TEL 2015. All underwent a peer-review selection: each paper was assessed by at least two different reviewers, from an international panel composed of about 60 members of 14 countries. The program of mis4TEL counts 11 contributions from diverse countries such as Colombia, Italy, Malaysia, Portugal and Spain. The quality of papers was on average good, with an acceptance rate of approximately 75%.

Last but not least, we would like to thank all the contributing authors, reviewers and sponsors (Telefónica Digital, Indra, INSA - Ingeniería de Software Avanzado S.A (IBM), IEEE Systems Man and Cybernetics Society Spain, AEPIA Asociación Española para la Inteligencia Artificial, APPIA Associação Portuguesa Para a Inteligência Artificial, CNRS Centre national de la recherché scientifique and STELLAR, as well as the members of the Program Committee, of the Organising Committee for their hard and highly valuable work. The work of all such people crucially contributed to the success of mis4TEL'15.

Tania Di Mascio
Rosella Gennari
Pierpaolo Vittorini
Fernando De la Prieta
(Editors)

Organization MIS4TEL 2015 – http://mis4tel.usal.es

Steering Committee

Fernando De la Prieta	University of Salamanca, Spain
Tania Di Mascio	University of L'Aquila, Italy
Rosella Gennari	Free University of Bozen-Bolzano, Italy
Pierpaolo Vittorini	University of L'Aquila, Italy

Program Committee

Silvana Aciar	Universidad Nacional de San Juan, Argentina
Yaxin Bi	University of Ulster, UK
Maria Bortoluzzi	Università degli Studi di Udine, Italy
Margherita Brondino	University of Verona, Italy
Edgardo Bucciarelli	University of Chieti-Pescara, Italy
Davide Carneiro	Universidade do Minho, Portugal
Gina Chianese	Free University of Bolzano/Bozen, Italy
Vincenza Cofini	University of L'Aquila, Italy
Juan Cruz-Benito	University of Salamanca, Spain
Paola D'Orazio	University Chieti Pescara, Italy
Giovanni De Gasperis	University of L'Aquila, Italy
Vincenzo Del Fatto	Free University of Bolzano/Bozen, Italy
Tania Di Mascio	University of L'Aquila, Italy
Gabriella Dodero	Free University of Bolzano, Italy
Liliana Dozza	Free University of Bolzano/Bozen, Italy
Néstor Darío Duque Méndez	Universidad Nacional de Colombia, Colombia
Richard Duro	Universidade da Coruna, Spain
Florentino Fdez-Riverola	University of Vigo, Spain
Manuel Jose Fernandez Iglesias	Universidade de Vigo, Spain
Rosella Gennari	Free U. of Bozen-Bolzano, Italy
Barbara Giacominelli	Verona University, Italy

Local Organising Committee

Javier Bajo   Technical University of Madrid, Spain
Juan F. De Paz   University of Salamanca, Spain
Sara Rodríguez   University of Salamanca, Spain
Fernando De la Prieta Pintado   University of Salamanca, Spain
Gabriel Villarrubia González   University of Salamanca, Spain
Javier Prieto Tejedor   University of Salamanca, Spain

Review Process

MIS4TEL welcomed the submission of application papers with preference to the topics listed in the call for papers.

All submitted papers followed a thorough review process. For each paper was be refereed by at least two international experts in the field from the scientific committee (and most of them by three experts) based on relevance, originality, significance, quality and clarity.

The review process took place from January 20, 2015 to February 18, 2015 and was carried out using the Easychair conference tool.

The review process granted that the papers consisted of original, relevant and previously unpublished sound research results related to any of the topics of the conference.

All the authors of those papers requiring modifications were required to upload a document stating the changes included in the paper according to the reviewers' recommendations. The documents, together with the final versions of the papers, were revised in detail by the scientific committee chairs.

Contents

Emotions and Inclusion in Co-design at School: Let's Measure Them!

Margherita Brondino[1], Gabriella Dodero[2], Rosella Gennari[2], Alessandra Melonio[2], Margherita Pasini[1], Daniela Raccanello[1], and Santina Torello[2]

[1] Faculty of Philosophy, Education and Psychology, University of Verona, Lungadige Porta Vittoria 17, 37129 Verona, Italy
{margherita.brondino,margherita.pasini, daniela.raccanello}@univr.it
[2] Faculty of Computer Science, Free University of Bozen-Bolzano, Piazza Domenicani 3, 39100 Bolzano, Italy
{gabriella.dodero,gennari,alessandra.melonio, santina.torello}@unibz.it

Abstract. Co-design with children comes with methods and techniques for creating technological products with children, such as video-game prototypes. When co-design takes place in schools, learners' involvement and enjoyment of co-design become crucial concerns for researchers. But how to measure emotions, more in general, and involvement in a co-design study with children? This paper presents a co-design study, run with a novel co-design method at school, for involving children in co-design groups and emotionally engaging them in producing game prototypes. It explains how emotional engagement and inclusion can be and were operationalized and measured in the co-design study, thereby providing feedback to co-design researchers interested in measuring the same constructs.

Keywords: Game design, co-design, gamification, cooperative learning, experience design, emotional engagement, inclusion, measure, children, schools.

1 Introduction and Related Work

Co-design, in the sense of [1], is a general approach to design that extends several others, such as participatory design and co-creation, involving potential users as collaborative designers. In principle, it can be used in any step of the design process for enabling "collective creativity", i.e., in the steps of (1) conceptualization of design solutions, (2) prototyping and development, (3) evaluation.

When users are children, specific methods are used. Notice that the word "method" is hereby used as in Fails et al. [2]: whereas a technique is a focused design activity used at varying points in the design process and with specific goals or outputs, a co-design method is a collection of techniques within an overall design philosophy. In the majority of co-design methods, children usually work in teams, often with adults. Team members have diverse functions: children become co-designers as "experts of their experience"; expert designers, with experience of the product under design, bring in their professional expertise. See [2] for an overview of co-design methods.

T. Di Mascio et al. (eds.), *Methodologies & Intelligent Systems for Technology Enhanced Learning*, Advances in Intelligent Systems and Computing 374, DOI: 10.1007/978-3-319-19632-9_1

When co-design studies take place at school, teachers can also be members of the co-design team, mainly as experts of education and of their school context. Co-design at school, albeit not new, has received increasing attention in recent years, e.g., [4]. This paper presents a co-design study at school for allowing children to create their own game prototypes. The study was conducted in Spring 2013. It used a novel co-design method for working at school with children organized in teams of 3–5 members, namely, *GAmified CO-design with COoperative learning* (GaCoCo). Its ideas were for the first time advanced in [5], and iteratively refined across 4 years of work with schools, so far involving circa 140 children and 8 teachers of different ages; an example of an iteration from a study to another is reported in [6].

The majority of co-design studies at school aim at including all children and engaging them emotionally in the experience [7]. However, to the best of our knowledge, we did not find research work that, so far, operationalized and measured both group inclusion and children's emotions in co-design, as in quantitative research [8]. This paper presents a way to do it, in the context of the 2013 study conducted with GaCoCo in classroom. Moreover, the paper investigates dependencies among the two constructs.

The GaCoCo co-design method is outlined in the first part of this paper, after introducing the necessary background information. The second part reports the GaCoCo study run in primary schools: it explains how emotions and inclusion were measured, and results concerning their dependencies.

2   GaCoCo Method

2.1   Introduction

GaCoCo is a co-design method with techniques that researchers can use to work with a school class, divided in groups, for the conceptualization and prototype development of games [9]: conceptualization techniques, for releasing parts of the so-called game design document; prototyping techniques, which leverage on existing co-design ones and specialize them to the release of a paper-based version of a game prototype with levels, in brief, a low-fidelity prototype. An example prototype by children is in Fig. 2.

Participants in GaCoCo have different functions. Children are the main designers, often working in groups of 3–5 members. Teachers take care of illustrating the daily work organization, and of moderating the class behavior. Researchers are usually two per class, with two different functions: one of expert (game) designer, who follows each group for providing rapid scaffolding feedback, and for conducting a formative evaluation of each group work at specific moments; the other, more experienced of child development studies, acts as passive observer, referred to as observer henceforth.

Moreover, GaCoCo comes with its own co-design philosophy, based on its unique blend of gamification and cooperative learning, whose contributions to GaCoCo are outlined in the remainder of this section.

2.2   Gamification Contributions

In its most common acceptation, gamification means properly using game-based elements, such as story lines and progression bars, for a non-game goal and in a non-game context in order to engage people, regarded as players, with positive emotions, e.g., see [10]. Following gamification principles, GaCoCo organizes co-design sessions as missions of a game via adequate gamified objects, and with a goal valuable for all co-designers. The relevance of the expert designer of the co-designed product fades though missions. According to their complexity, GaCoCo chunks missions into small progressive challenges, disclosed when needed with clear rules, of which the first challenge is easy to take up by all learners. Progression maps or completion rewards, which are contingent to co-design, are used by GaCoCo for conveying the idea of growth. Rewards can be tangible or not. Examples of GaCoCo tangible rewards are (fake) coins, earned on completion of co-design challenges; positive failure feedback in case of errors in challenges is an example of a non-tangible reward.

A co-design context that invites children's free exploration and choice, like an unexplored game world, can even more tangibly promote a sense of autonomy and control over their co-design work. A simple example is as follows: on completion of a mission, children are invited to choose one among different completion rewards, which they may also customize, and to use the reward for their next co-design mission. In this manner they gain the feeling that their actions have a tangible effect on their co-design work, and that they are in control of choosing parts of this. See the shop for buying co-design objects as rewards, using coins, at the end of a co-design mission, in Fig. 2.

Relatedness needs are also important components of games. Gamification of co-design contexts, however, should be done fostering cooperation, so as to be faithful to the co-design partnership principle, e.g., without increasing competition within groups. Such a constraint can be met by providing rewards that mildly favor only intergroup competition so as to promote "intra-group positive interdependence" and cooperation [11], and still satisfy relatedness needs. Progression maps can also be used to connect with others and satisfy relatedness needs. Shared maps can show other learners that a group or an individual could overcome a co-design mission, and are available for sharing their co-design experience and acquired expertise.

Fig. 1. Signaling disk (black and brown) and scepter (yellow) usage (left), and a screenshot of a progression map for challenges (right)

Fig. 2. The shop in the classroom for buying objects for game prototypes or the expert help card (left), and a game prototype (right)

2.3 Cooperative Learning Contributions

Co-designers need to develop a sense of partnership, which is crucial to manage complex group dynamics when co-designing with children [12]. Cooperative learning is an instructional methodology that can help in organizing and managing group dynamics in classroom. How this can be done varies according to the chosen cooperative learning model. GaCoCo adopts the Complex Instruction model [13]. Heterogeneity, in this model, becomes a growth opportunity at the cognitive and social levels. The strategies for organizing the work of heterogeneous groups, the rules and roles for children are all important cooperative learning means that GaCoCo acquires and adapts from the Complex Instruction model to the end of co-design, making them tangible via gamified objects, as those in the above examples. The following part exemplifies how.

There are a number of *strategies* for organizing group work in cooperative learning; examples are co-op co-op, for organizing work for groups of 4, and gallery tour, for sharing opinions or products across groups [14]. GaCoCo employed several of them, e.g., gallery tour. In GaCoCo studies, gallery tours were used for sharing opinions on game prototypes, and gamified via specific objects, e.g., banknotes for voting the preferred prototype. The voting box is highlighted at the bottom of Fig. 2 with blue.

Roles facilitate the management and working of the group as a team. Each member of a group is assigned a role. In GaCoCo, roles are not fixed but rotate among members so that all train different skills. The group ambassador is an example of a cooperative learning role for children adopted in GaCoCo studies. Ambassadors ask for clarifications; more generally, they are responsible for exchanging information with the teacher and design expert. Roles can be conveyed via specific objects. For instance, in GaCoCo studies, ambassadors had their gamified object, the expert card, for asking the help of the game expert at critical points. The card is highlighted in green in Fig. 2.

Besides strategies and roles, and in support of them, cooperative learning considers a set of *rules* necessary for working in group and including all. Rules are concerned with social skills such as reciprocal listening and respect of different views. Examples of cooperative learning rules that GaCoCo employs are taking turns in voicing opinions, and reconciling different views, e.g., concerning game concepts or prototypes. GaCoCo studies made such rules clear and easy to recognize by using ad-hoc gamified objects. For example, each group was endowed with a scepter for organizing turns in speaking. Each child could vote on different views by drawing smileys on signaling-disks. See a scepter and a signaling-disk in Fig. 1.

3 Study Design

The GaCoCo study hereby reported was in two primary schools, involving a class of 15 8–9 year olds, and a class of 20 9–10 year olds (females 59%). It was organized as an empirical one along 3 main activities: pre-activity, main activity, post-activity. Only the relevant details of the study are reported below.

3.1 Aims and Data Collection Instruments

Our general aim was to investigate relationships in GaCoCo between emotional engagement, which is claimed to be activated by gamification and to contribute to learners' performance at school [15], and social inclusion, assumed to be fostered by a cooperative learning approach. We operationalized emotional engagement in terms of achievement emotions—linked to learning activities or outcomes—experienced by children as co-designers at school. The reference theoretical framework is Pekrun's control-value model [16]. We operationalized social inclusion using a specific sociometric status indicator, that is, peer reciprocity [17].

3.2 Hypotheses

We hypothesized positive links between positive emotions and peer reciprocity, and negative links between negative emotions and peer reciprocity, extending current literature [17]. We also explored whether achievement emotions played a partial or total mediating role between pre-intervention peer reciprocity and post-intervention peer reciprocity; in other terms, we examined how important emotions were in determining changes in the sociometric status.

3.3 Pre-activity: Training

During the pre-activity, designers organized a meeting with the school dean and interested teachers in each school in order to explain and discuss the project. A week before the main activity, a workshop for teachers, lasting circa 6 hours, was organized and a more focused training was thus performed. For instance, during the workshop, the protocol of each mission was explained by expert designers, and so were the main ideas of gamification and game design. Teachers worked in group and experimented the protocol for children by prototyping games themselves. After the workshop, teachers were asked to create heterogeneous groups of children in terms of learning and social skills. Also children were trained to game design principles. This training lasted circa 20 minutes. Moreover, teachers administered children a sociometric task to obtain a peer reciprocity score [17], which included three questions asking each child to choose one classmate in relation to three contexts. We calculated a peer reciprocity score as mean of the total number of reciprocated choices obtained by each child in the three contexts, divided by the number of children included in each class. This is referred to as the pre-reciprocity score.

3.4   Main Activity: GaCoCo Co-design

The main activity had six sessions, each of which was gamified and presented as a mission, as explained in Subs. 2.2, and organized with cooperative learning rules, roles and strategies, outlined in Subs. 2.3. The last mission was run in the university premises, whereas all the others were run in classroom. This paper focusses on the five missions in classroom.

Each mission at school took a different day, lasting circa 2 hours and a half. Missions were done every week, the same day of the week. All missions were organized linearly with challenges, different from mission to mission. Each challenge came with its specific conceptualization and prototyping techniques for releasing parts of groups' game design documents and prototypes. In the second mission, each group released the high-level concept document; then they used it for prototyping their game characters and objects. In the third and fourth missions, each group conceptualized and prototyped two game levels, working first in pairs and then sharing results in group. In the fifth mission, each group firstly conceptualized and secondly prototyped the passage between levels, thereby releasing a single game prototype. Groups also tested their presentation of their game in view of the sixth mission at university.

After each mission, children were administered the Graduated Achievement Emotion Set (GR-AES) of [18], a verbal-pictorial instrument currently under development, aiming at assessing the intensity of ten achievement emotions. Children were asked to rate each emotion with a 5-point Liker-type scale with 5 faces corresponding to different levels of emotion intensity (from 1, for "not at all", to 5, for "extremely"). Our study considered the following achievement emotions: three positive-activating emotions (enjoyment, hope, pride), two positive de-activating emotions (relief, relaxation), three negative-activating emotions (anxiety, anger, shame), two negative de-activating emotions (boredom, sadness).

3.5   Post-activity: Debriefing

In the post-activity, debriefing interviews with children were run by teachers with the help of researchers. The same sociometric task of the pre-activity was administered by teachers to children, so as to obtain the post-reciprocity score.

4 Results and Discussion

In statistics, path analysis is used to describe the directed dependencies among a set of variables. A series of path analyses was carried out to study the relationships between achievement emotions, measured at the end of each mission at school, and peer reciprocity scores, measured both pre and post activity. We computed a composite score of emotion intensity across missions calculating the mean value for each emotion. We included the pre-reciprocity score as predictor of achievement emotions, in turn predictors of the post-reciprocity score, to verify the mediating role of achievement emotions. Our hypothesis was confirmed for two emotions, which are among the most salient in learning contexts. The pre-reciprocity score positively predicted enjoyment ($\beta = .35$, $p < .05$), that in turn positively predicted the post-reciprocity score ($\beta = .34, p < .05$), whose explained variance was 11%. The mediating role of enjoyment was partial (indirect effect $= .12, p < .05$) because the path between pre-reciprocity score and post-reciprocity score was statistically significant ($\beta = .67, p < .001$). In turn, anxiety negatively predicted the post-reciprocity score ($\beta = -.37, p < .05$), and it explained the 13% of its variance.

Summing up, we found that the emotions of enjoyment and anxiety were crucial to influence learners' sociometric status. Future studies including a larger number of participants could examine the generalizability of such effects. On the whole, our results suggest that monitoring achievement emotions within a co-design study assumes particular relevance in order to track and possibly stir changes in group dynamics in co-design learning contexts.

5 Conclusions

This paper briefly presents a co-design study in primary schools for realizing paper-based prototypes of games for children, by children, run in Spring 2013. The study used the GaCoCo co-design method and techniques, organizing co-design work for groups of children with cooperative learning, and presenting co-design sessions as missions with specific gamified material. Gamification and cooperative learning were used in the study for soliciting children's emotional engagement and for including all children in the co-design work. Such constructs (emotional engagement and inclusion) are only qualitatively assessed in general in the co-design literature. The paper, for the first time, presents a specific way for operationalizing and hence measuring, using quantitative data collection instruments, in the context of the 2013 study: sociometric status was operationalized considering peer-reciprocity, whereas emotions were operationalized in terms of achievement emotions.

The paper concludes presenting the results of a series of statistical path analyses for studying the relationships between achievement emotions and peer-reciprocity scores. According to the available results, enjoyment played a significant role in improving learners' sociometric status, while anxiety was associated with a significant deterioration of the sociometric status. Such data support the relevance of fostering positive emotions and reducing negative emotions for favoring inclusion and properly managing group dynamics across a co-design journey in a school context.

References

1. Sanders, E.B., Stappers, P.J.: Co-creation and the New Landscapes of Design. CoDesign: International Journal of CoCreation in Design and the Arts 4(1), 5–18 (2008)
2. Fails, J.A., Guha, M.L., Druin, A.: Methods and Techniques for Involving Children in the Design of New Technology for Children. Now Publishers Inc., Hanover (2013)
3. Sleeswijk Visser, F., van der Lugt, R., Stappers, P.: Participatory Design Needs Participatory Communication. In: Proceedings of 9th European Conference on Creativity and Innovation, pp. 173–195 (2005)
4. Vaajakallio, K., Lee, J., Mattelmäki, T.: "It Has to Be a Group Work!": Co-design With Children. In: Proceedings of IDC 2009, pp. 246–249. ACM (2009)
5. Dodero, G., Gennari, R., Melonio, A., Torello, S.: Gamified Co-design with Cooperative Learning. In: CHI 2014 Extended Abstracts on Human Factors in Computing Systems, CHI EA 2014, pp. 707–718. ACM, New York (2014)
6. Dodero, G., Gennari, R., Melonio, A., Torello, S.: Towards Tangible Gamified Co-design at School: Two Studies in Primary Schools. In: Proceedings of the First ACM SIGCHI Annual Symposium on Computer-human Interaction in Play, CHI PLAY 2014, pp. 77–86. ACM, New York (2014)
7. Mazzone, E.: Designing with Children: Reflections on Effective Involvement of Children in the Interaction Design Process. Phd thesis, University of Central Lancashire (2012)
8. Mueller, C.: Conceptualization, Operationalization, and Measurement. In: The SAGE Encyclopedia of Social Science Research Methods. Sage Publications, Thousand Oaks (2004)
9. Adams, E.: Fundamentals of Game Design, 3rd edn. Pearson, Allyn and Bacon (2013)
10. Kapp, K.M.: The Gamification of Learning and Instruction. Wiley (2012)
11. Romero, M., Usart, M., Ott, M., Earp, J.: Learning through Playing for or against Each Other? Promoting Collaborative Learning in Digital Game Based Learning. In: Proceedings of ECIS, p. 93 (2012)
12. Van Mechelen, M., Gielen, M.: vanden Abeele, V., Laenen, A., Zaman, B.: Exploring challenging group dynamics in participatory design with children. In: Proceedings of the 2014 Conference on Interaction Design and Children, IDC 2014, pp. 269–272. ACM, New York (2014)
13. Cohen, E.: Making Cooperative Learning Equitable. Educational Leadership (1998)
14. Kagan, M., Kagan, S.: Cooperative Learning. Kagan Cooperative Learning (1992)
15. Fredricks, J., McColskey, W.: Measuring Student Engagement in Upper Elementary Through High School: A Description of 21 Instruments. Technical report, Institute of Education Sciences: U.S. Department of Education (2011)
16. Pekrun, R., Perry, P.: Control-value Theory of Achievement Emotions. In: International Handbook of Emotions in Education. Taylor and Francis (2014)
17. Santos, A., Vaughn, B., Peceguina, I., Daniel, J.: Longitudinal Stability of Social Competence Indicators in a Portuguese Sample: Q-sort Profiles of Social Competence, Measures of Social Engagement, and Peer Sociometric Acceptance. Developmental Psychology 50(3), 968–978 (2014)
18. Raccanello, D., Bianchetti, C.: Achievement Emotions in Technology Enhanced Learning: Development and Validation of Self-report Instruments in the Italian Context. Interaction Design and Architecture (submitted)

Emotional Adaptive Platform for Learning

Ana Raquel Faria[1], Ana Almeida[1], Constantino Martins[1], and Ramiro Gonçalves[2]

[1] GECAD - Knowledge Engineering and Decision Support
Research Center Institute of Engineering – Polytechnic of Porto (ISEP/IPP), Portugal
{arf,amn,acm}@isepp.ipp.pt
[2] Universidade de Trás-os-Montes e Alto Douro, Portugal
ramiro@utad.pt

Abstract. The aim of this paper is to present a new approach in user modeling process that use learning and cognitive styles and student emotional state to adapt the user interface, learning content and context. The model is based on a constructivist approach, assessing the user knowledge and presenting contents and activities adapted to the emotional characteristics, learning and cognitive styles of the student. The intelligent behavior of such platform depends on the existence of a tentative description of the student – the student model. The contents of this model and emotional state of the student are used by a domain and interaction model to select the most suitable response to student actions.

Keywords: Learning Styles, Student Modeling, Adaptive Systems, Affective Computing.

1   Introduction

The interaction-reaction between the teacher and student during a "traditional" classroom has a significant effect (positive or negative) in the effectiveness of the learning process [1]. In a learning system, this interaction is limited, for example most of the time, the feedback send by the students or teachers are usually worked very slowly which contributes to the frustration and boredom of the students. This interaction does not trigger any real time adaptation to the content or context presentation, in terms of flow, organization or difficulty. Therefore we propose new approaches that take advantage of the student emotional state, and simulate the teacher-student interaction and real time adjustment of course parameters (flow, content and context, organization or difficulty) in order to recapture the student attention. Learning process personalization, mainly when referring to learning systems can be a good way of improving learning effectiveness. This means adapting al the learning process, namely, domain contents, content presentation and user interface accordingly students' specific profile (which includes learning style, psychological and personality profile), students' knowledge in the domain application, and also emotional state.

2   Affective Learning and Models

There are numerous ways of learning. In 1956 Bloom [2], identified three domains of educational activities: the Cognitive: mental skills (Knowledge), the Affective:

© Springer International Publishing Switzerland 2015
T. Di Mascio et al. (eds.), *Methodologies & Intelligent Systems for Technology Enhanced Learning,*
Advances in Intelligent Systems and Computing 374, DOI: 10.1007/978-3-319-19632-9_2

growth in feelings or emotional areas (Attitude) and the Psychomotor: manual or physical skills (Skills). The blend of all the domains influences the way that a person learns and the way rational decisions are made. So what is Affective Learning? According to recent definitions, "Affective learning involves the melding of thinking and feeling in how people learn. Importance is placed on social learning environments for knowledge construction and application wherein deeper awareness and understanding of the role played by mental dispositions in how a person views, engages, and values learning can result in better understanding and use of knowledge and skills. Learning outcomes are focused on enculturation of norms, values, skillful practices, and dispositions for lifelong learning." [3]. The goal of the use of a model is to understand how the emotions are evolving in the learning process. So learning systems can be developed in order to recognize and responded appropriately to an emotional state. In 2004 Picard indicated that the theories that influenced the learning process need to be verified and further developed. Most of the models have been developed like a model for pervasive e-Learning platform [4] had as a starting point models like Russell's Circumplex model [5] to describe user's emotion space and Kort's learning spiral model [6] to explore the affective evolution during learning process.

3   Learning Styles

Generally, Learning Styles are understood as something that intent to define models of how a person learns. Generally it is understood that each person has a Learning Style different and preferred with the objective of achieving better results. Some case that teachers should assess the learning styles of their students and adapt their classroom methods to best fit each student's learning style [7] [8]. VARK Strategies is a questionnaire that provides users with a profile of their learning preferences. These preferences are about the ways that they want to access and select content. These strategies describe three basic learning preferences: Visual learning (learn by seeing); Auditory learning (learn by hearing) and Kinesthetic or practical learning (learn by doing) [9]. There are different learning styles models (based on different psychological theories) such as for example models based on (Kolb 2005): Personality [10], Information processing approach [11], Social Interaction [12], Multidimensional factors [10].

4   Adaptive System and Student Model

An Adaptive System (AS) builds a model of the objectives, preferences and knowledge of each user and uses it, dynamically, through the Domain Model and the Interaction Model, to adapt its contents, navigation and interface to the user needs [13]. In Educational Adaptive Systems, the emphasis is placed on students' knowledge in the domain application and learning style, in order to allow them to reach the learning objectives proposed in their training [13].

In generic Adaptive Systems (AS), the User Model allows changing several aspects of the system, in reply to certain characteristics (given or inferred) of the user [9]. These characteristics represent the knowledge and preferences that the system assumes that the user (individual, group of users or no human user) has. In Educational AS, the UM (or Student Model) has increased relevance: when the student reaches the objectives of the course, the system must be able to re-adapt, for example, to his knowledge [9]. A Student Model (SM) includes the Domain Dependent Data (DDD) and the Domain Independent Data (DID). The Domain Independent Data (DID) are composed of two elements: the Psychological Model and the Generic Model of the Student Profile, with an explicit figure [15].

5 Proposed Prototype

In order to prove that affective learning can have influence in the learning process. A prototype was developed, a learning platform that takes into account the emotional aspect, the learning style and the personality traits, adapting the course to the student objectives. The architecture is divided in four major models: the Student Model, the Emotion Model, the Emotive Pedagogical Model and the Application Model. The Figure 1 shows the architecture of the prototype.

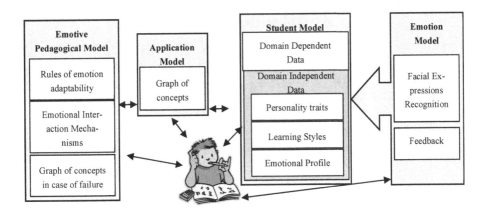

Fig. 1. Architecture

5.1 Student Model

The student model consists in retrieving the student information and characteristics. This can include personal information (name, email, telephone, etc.), demographic data (gender, race, age, etc.), knowledge, deficiencies, learning styles, emotion profile, personality traits, among others [9]. This information is useful to better adapt the prototype to the student needs. The approach used to build the Student model is the Stereotype Model with the overlay model for the knowledge representation of the student. The representation of the stereotype is hierarchical. Stereotypes for users with different knowledge have been used to adapt information, interface, scenario, goals and plans [9]. For the definition of the Learning Styles of the student we are using the Kolb Learning Styles Matrix. Concerning that and the objective of Domain Dependent Data, users aptitude and assessments result will be monitoring. The Initialization phase setups the user model and other application of features extraction setups. For the first time the user has to answer a questionnaire/surveying that could be, partly answered from existing databases. For example, if the learning platform is situated in a school, the information from the registration and history of the student can be used. The questionnaire allows the system to start a new user model based on the user answers and completed with the collection of data from the features extraction application. When designing a user model several questions have to be ask, like who is being modelled; what aspects of the user are being modelled; how is the model to be initially acquired; how will it be maintained; and how will it be used [16]. Who is being modelled, the first thing we have to establish is if we are modelling single users or a group of users and if the model as short or long term life. In a model which the information of individual user will persist over time, we have to take into account that information may change or grow. Also, what aspects of the user are being modelled, if this information should represent generic facts that are common for the majority of users or particular facts that portraits the user knowledge, beliefs and general understanding others, subjects or things of the user or a combination of both.

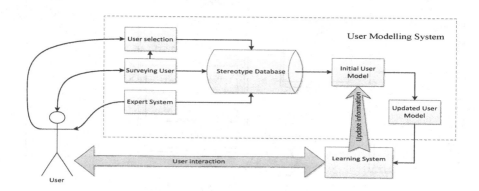

Fig. 2. User Modelling System

The user model can be initially acquired by selecting one or more of the existing stereotypes existing in the system [17]. The selection of a suitable stereotype can be achieved by a number of techniques. These techniques include: the user being able to decide which model wants to use, surveying de user and from his answers choosing the most appropriate model or even having an expert system to analyze the user and selecting one of the models. Also the initial model can be chosen from data collected in existing database, for instance from an LDAP database. As soon as the initial user model is selected it can be updated, maintained with particular data about the user.

5.2   Emotion Model

Facial Expression Recognition allows video analyses of frames in order to recognize an emotion. This is done by making use of an API called ReKognition[1]. This API allows detection of the face, eyes, nose and mouth and if the eyes and mouth are open or close. In addition specifies the gender of the individual, an estimate of age and emotion. In each moment a group of three emotions are captured and each one has tailing number that shows the confidence. The feedback consists in a series of questions at the end of each subject. That aims to discover the impression and rate of the subject, the student has learned. Questions like: "How would you rate this subject?" ☆☆☆☆☆ or "What emotion are you feeling?"

5.3   Application Model

The application model is compose by a series of elements contain different subjects. The subject consist in a number steps that the student has to pass in order to complete is learning program. Some of these steps are optional others are not. First step is the subject is placement test (PT) that can be optional. This is designed to give students and teachers a quick way of assessing the approximate level of a student's knowledge. The result of the PT is percentage PTs that is added to the knowledge (Ks) of the student on a subject and places the student on one of five levels of knowledge. $kpt = \sum_{i=0}^{5} exercise_i$ If the PT is not performed Ksp will equal to zero and the student will start with any level of knowledge. The SC contains the subject explanation. The explanation of the subject depends on the stereotype (Student Model and the combination of learning styles, personality dimensions and motivations). Each explanation will have a practice exercise. These exercises will allow the students to gain points to do the final test of the subject. The student needs to get 80% on the TotalKsc to do the test. The ST is the evaluation of the subject learned. The will give value kst to the knowledge of the student on the subject $kst = \sum_{i=0}^{5} exercise_i$ only if the kst is higher than 50% the student as successful completed the subject. In this case the values of the ksp and kst are compared to see if there was an effective improvement on the student knowledge.

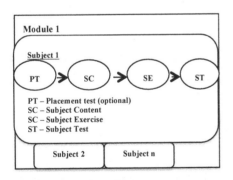

Fig. 3. Concepts Graph

---

[1] "Orbe.us | ReKognition - Welcome to Rekognition.com." [Online]. Available: http://rekognition.com/index.php/demo/face [Accessed: 26-Jul-2014].

5.4   Emotive Pedagogical Model

The Emotive Pedagogical Model is compose by three sub models; the Rules of emotion Adaptability, the emotional interaction mechanisms and the Graph of concepts in case of failure. In each step of a subject the emotional profile is update by facial expression recognition software. The emotional expression in order to establish an emotional profile and with this information make some adjustments to content of the application or trigger an emotional motivation action or emotional interaction. The Rules of emotion Adaptability consist in the way the subject content is present. The subject content is presented according the learning style and personality of the student. For a determined personality there is a manner a how the information and exercises can be formulated and according to a specific learning style. The information and exercises can be adapted to the student requirements. The Emotional Interaction Mechanisms consist in trigger an emotion interaction, when is captured an emotion that needs to be contradicted in order to facilitate the learning process. The some of the emotions to be contradicted are: anger, sadness, confusion and disgust. The interaction can depend on the personality and on the learning style of the student. The Emotion interaction Mechanisms is composed by the Action Proposer. The Action Proposer recommends actions base on the emotion capture and user characteristics/ personality and model.

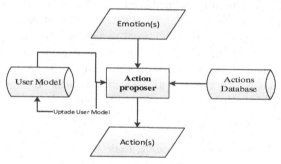

Fig. 4. Action proposer

For example if a specific emotion captured that emotion can be contradicted, induce to produce a different emotion or sustained this is done by proposing several actions. In the following table show an example how this process can occur. The emotion capture is sadness and base on the user model a number of actions are triggered to be chosen.

Table 1. Emotion/Actions

Emotion	User Model	Actions
Sadness	Visual Student	Action1: Show images that can remind of happy things, like comedies
		Action2: Show tragic plays or pictures
	Aural Student	Action3: Play sounds or music with an up-beat rhythm.
		Action4: Play sounds or music with a down-beat rhythm.
	Verbal Student	Action5: Use phases that contain words such as 'new', 'exciting' and 'wonderful' to trigger a feel good sensation.
		Action6: Show depressing new.

The graph of concepts in case of failure is the steps to be taken when fail to pass a subject. To pass the module the entire subjects must be completed, and only with a subject completed you can pass to the next. Inside of a subject the student has to complete e the placement test, the subject content plus exercises with a grade equal or higher than 80% and the subject test with approval with a grade higher than 50% to complete the subject. In case of failure it has to go back to the subject content and repeat all the steps.

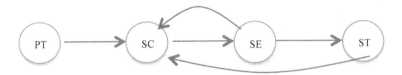

Fig. 5. Graph of concepts in case of failure

6   Conclusion

The aim of this paper is to near the gap between a student and his learning platform in order to improve efficiency of the learning process. We propose of a new approach for an emotional adaptive learning system to try to solve the problem. The architecture proposed for this model is composed of 4 major models: the Student Model, Emotional Model, the Application Model and the Emotive Pedagogical Model. This model will try to capture the emotional state of the student and together with his learning style and personality profile, will adapt the learning context to the learning requirements of the student. Based on this model a prototype was developed and in order to evaluate the prototype it has chosen a course of the first year of Polytechnic school of engineering. Emotion shapes almost all modes of human communication, from facial expression, gestures, posture, tone of voice, choice of words, respiration, and skin temperature, to many others. Today advances in recognition allow creation of new ways of human-computer interaction, making this interaction more usable and intuitive. To near the gap between a student and his learning system in order to improve efficiency of the learning process, it is addressed in this paper, under the proposed approach for an adaptive learning system. This model will try to capture the emotional state of the student and together with his learning style and cognitive profile, will adapt the learning context to the learning needs of student. The next steps in the development of the platform will be to explore natural languages sentences and clustering algorithms to identify difficulties among the students through the analysis of the message dialog student-student or student-teacher. This feature will turn the platform into a full collaborative platform, where students may share their difficulties and to get appropriate feedback. In order to evaluate the system a set of courses in one Polytechnic school will be used. Some research issues addressed by this undergoing work, still remaining to be answered, what are the teaching-learning contexts of the considered emotions? How do they relate to student learning? What are the practical implications for teaching and learning? Obtaining the answers to these questions implies further work including exhaustive analysis of case studies in order to compare results under different contexts.

Acknowledgements. This work is supported by FEDER Funds through the "Programa Operacional Factores de Competitividade - COMPETE" program and by National Funds through FCT "Fundação para a Ciência e a Tecnologia" under the project: FCOMP-01-0124-FEDER- PEst-OE/EEI/UI0760/2014.

References

1. Luan, W.S., Bakar, K.A.: The Shift in the Role of Teachers in the Learning Process 7(2), 33–41 (2008)
2. Bloom, B.S.: Taxonomy of Educational Objectives, Handbook 1: Cognitive Domain. Addison Wesley Publishing Company (1956)
3. Stricker, A.G.: Why Affective Learning in a Situated Place Matters for the Millennial Generation (2009)
4. Shen, L., Wang, M., Shen, R.: Affective e-Learning: Using 'Emotional' Data to Improve Learning in Pervasive Learning Environment Related Work and the Pervasive e-Learning Platform. Learning 12, 176–189 (2009)
5. Russell, J.A.: A Circumplex Model of Affect, pp. 1161–1178 (1980)
6. Kort, B., Reilly, R., Picard, R.W.: An affective model of interplay between emotions and learning: reengineering educational pedagogy-building a learning companion. In: Proc. IEEE Int. Conf. Adv. Learn. Technol., pp. 43–46 (2001)
7. Kolb, A.Y.: Learning styles and learning spaces: Enhancing experiential learning in higher education. J. Acad. Manag. Learn. Educ. 4(2), 193–212 (2005)
8. Stash, P., Cristea, N., De Bra, A.: Explicit Intelligence in Adaptive Hypermedia: Generic Adaptation Languages for Learning Preferences and Styles. In: Proc. Work. CIAH HT 2005, pp. 75–84 (2005)
9. Martins, C., Couto, P., Fernandes, M., Bastos, C., Lobo, C., Faria, L., Carrapatoso, E.: PCMAT – Mathematics Collaborative Learning Platform. In: Pérez, J.B., Corchado, J.M., Moreno, M.N., Julián, V., Mathieu, P., Canada-Bago, J., Ortega, A., Caballero, A.F. (eds.) Highlights in PAAMS. AISC, vol. 89, pp. 93–100. Springer, Heidelberg (2011)
10. Ritu, M., Sugata, D.: Learning Styles and Perceptions of Self. Int. Educ. J. 1(1), 61–71 (1999)
11. Schmeck, R.R.: Learning styles of college students. In: Dillon, R., Schmeck, R. (eds.) Individ. Differ. Cogn., pp. 233–279. Acad. Press, New York (1983)
12. Reichmann, A.F., Grasha, S.W.: A rational approach to developing and assessing the construct validity of a student learning style scale instrument. J. Psychol. 87(2), 213–223 (1974)
13. Martins, A.C., Faria, L., De Carvalho, C.V.: User Modeling in Adaptive Hypermedia Educational Systems. Educ. Technol. Soc. 11, 194–207 (2008)
14. Martins, C., Azevedo, I., Carvalho, C.: The use of an Adaptive Hypermedia Learning System to support a new pedagogical model. In: The 5th IEEE International Conference on Advanced Learning Tecnologies – ICALT 2005 (2005)
15. Kobsa, A.: User Modeling: Recent Work, Prospects and Hazards. In: Adapt. User Interfaces Princ. Elsevier, Amsterdam (1993)
16. Finin, T., Orager, D.: GUMS 1: A General User Modeling System. In: Proc. Strateg. Comput. - Nat. Lang. Work., pp. 224–230 (1986)
17. Rich, E.: User Modeling via Stereotypes. Cogn. Sci. 3(4), 329–354 (1979)

On-line Assessment of Pride and Shame: Relationships with Cognitive Dimensions in University Students

Daniela Raccanello, Margherita Brondino, and Margherita Pasini

Department of Philosophy, Education and Psychology,
University of Verona (Italy), Lungadige Porta Vittoria, 17, 37129 Verona, Italy
{daniela.raccanello,margherita.brondino,
margherita.pasini}@univr.it

Abstract. We investigated some psychometric properties of an instrument developed to evaluate achievement emotions, focusing on pride and shame, with an on-line assessment. These emotions, as two moral emotions partially neglected by the literature, can also be categorized as achievement emotions, because of their salience within everyday academic life. The participants were 83 Italian university students. During their first year of university, we measured text comprehension; during their third year, we measured pride and shame (with the on-line version of the Brief-Achievement Emotions Questionnaire) and academic performance. Path analyses indicated the partial mediating role of the two emotions between text comprehension and academic performance, differing in study and test settings. A repeated-measure analysis of variance revealed higher levels of pride compared to shame, and of emotions associated with the evaluative setting. Such findings support the relevance of assessing emotions related to learning, for example monitoring them continuously through technological instruments.

Keywords: Achievement emotions, Cognitive abilities, Learning Environment Design, Emotions for Technology Enhanced Learning.

1   Introduction

Nowadays, there is an increasing diffusion of advanced learning technologies within everyday contexts, included school and academic settings [5]. As for traditional learning environments, a complex pattern of interrelations between cognitive, affective, and motivational constructs is linked to learning outcomes [7]. However, the psychological literature has only recently begun to investigate such interdependence as regards technological environments, focusing specifically on affect. For example, some research studies have demonstrated that individual differences in achievement emotions influence students' learning decisions related to the preference for using online or face-to-face learning modes [20]. Other researchers have found that only a small number of emotions is salient for learning in technology-based environments, specifically engagement and flow among positive emotions, and boredom, frustration, and confusion among negative emotions [5]. Yet, relationships between emotions and

© Springer International Publishing Switzerland 2015
T. Di Mascio et al. (eds.), *Methodologies & Intelligent Systems for Technology Enhanced Learning,*
Advances in Intelligent Systems and Computing 374, DOI: 10.1007/978-3-319-19632-9_3

other constructs are complex and some contradictory results have been found. For example, while some research studies documented the links between adaptive and maladaptive motivational patterns and positive and negative emotions, respectively, others revealed that for motivated students anger can be associated with persistence in order to master dense technical materials [5].

One way to foster the investigation of emotions within technological learning setting is to develop methodological instruments having the potentiality to be used in such contexts respecting reliability and validity standards, both creating new instruments and adapting traditional ones [18]. In such contexts, self-report measures concurrent with learning activities are indeed among the most frequently used methodologies, enabling to track also moment-to-moment affective states [5]. While recently different authors have proposed new ways to evaluate emotions in technological learning (e.g., Affective AutoTutor, trialogs with agents, or Crystal Islands), in order to track and respond adequately to the fleeing emotions associated with learning [5], also the adaptation of classical self-report measures has its advantages, such as the possibility to compare face-to-face and technological learning environments.

In this line, we developed a self-report measure, the Brief Achievement Emotions Questionnaire, BR-AEQ [1, 15, 17, 18], as a brief version of the Achievement Emotions Questionnaire, AEQ, by Pekrun and colleagues [12]. Such instrument was previously used with a face-to-face administration mode [1, 15, 17, 18], but its characteristics make it particularly suitable also for Technology Enhanced Learning (TEL) environments, in which limitations related to time constraints or to the need to assist individually people might pose methodological problems. Specifically, the questionnaire is characterized by clear language, brief administration, and simple answer format, specifically multiple-choice questions, which have been demonstrated as associated to low anxiety [14]. Thus, in light of the pervasive use of technology within educational contexts, investigating its psychometric properties, in terms of reliability and validity issues, is of primary relevance to open the door to its use for the design of TEL products, in order to assess affective dimensions, before, during, or after learning.

2   Background

In this work, we focused on pride and shame, as moral emotions 'linked to the interests or welfare either of society as a whole or at least of persons other than the judge or agent' [6, p. 853] poorly investigated in the psychological literature [21]. Such emotions are defined as self-conscious emotions, requiring the ability to think about the self, which emerges at the end of the second year, and implying a consequent immediate punishment or reinforcement to one's own behaviours [19]. They also play a key role in human adaptation to the social environment, modulating the relationships between moral standards and decisions on behaviours [8, 19]. In light of such strict interconnections between moral emotions and people's behaviours, pride and shame should be of primary focus within educational contexts. In these environments, supporting teachers' and students' emotional awareness could be a fruitful way to promote their general well-being, as a construct 'that encompasses all the dimensions (social, emotional, physical, cognitive and spiritual)' [9, p. 23] and consequently enables to plan intervention programs focusing on a 'whole person' learning [2, 9].

However, the two emotions differ for aspects such as valence–usually pride is considered as a positive emotion, while shame as a negative emotion, at least in Western cultures–and their antecedents. Pride derives from self-approval associated with personal success, resulting from achievement-related behaviours; shame is due to the perception of failures or transgressions, usually recognized by other people, with the focus on negative elements of the self rather than on negative behaviours [19, 21].

According to the control-value theory [11, 13], pride and shame are also categorized as achievement emotions because of their salience within everyday academic life. Achievement emotions are focused on learning activities and outcomes, and distinguished by valence (positive, negative emotions) and activation (activating, deactivating emotions). The model focuses on both antecedents and consequences of emotions [11, 13]. It posits that they are elicited by individual and social determinants, among which a key role would be played by individuals' beliefs on control, such as self-efficacy, and on values, in turn influenced by cultural values about the relevance of achievement. They would also have a range of consequences at different levels, involving cognitive, motivational, and behavioural aspects. According to this model, pride would be associated to positive academic performance, while the association between negative-activating emotions such as shame and performance would be more complex, even if the long-term disadvantageous effects of negative emotions on well-being and health are well acknowledged [14]. The model also assumes that achievement emotions would be organized in a context-specific way. This assumption has been verified empirically, operationalizing the context in terms of school domain, for emotions such as enjoyment, pride, anxiety, anger, and boredom; also the higher relative impact of domain compared to setting on emotions such as pride has been revealed [e.g., 3, 4, 16]. Finally, while knowledge about context-specificity of pride and shame, defining context in terms of settings, has been demonstrated with face-to-face assessment [17], data related to on-line assessment modes are still lacking. Awareness concerning the degree to which students' perceptions, beliefs, or emotional experiences vary in different contexts is of primary relevance for instructional practices, because it could help to organize technological environments in flexible ways.

2.1 The Present Study

Our general aim was to investigate some psychometric properties of an instrument developed to assess achievement emotions in face-to-face environments, used in technological environments with university students, focusing specifically on two emotions neglected by the psychological literature, pride and shame [21]. Considering context-specificity of emotions related to learning [11, 13], all the emotions were contextualize in two different settings, studying and taking tests related to university courses. First, we examined the reliability of the instrument. Second, we investigated its external validity, considering its mediating role between previous cognitive abilities and subsequent academic performance. On the basis of the literature [11, 13], we expected that performance was predicted positively by pride and negatively by shame, for both settings. Finally, we explored differences in the mean values of the two emotions, expecting the positive emotion to be more intense than the negative emotion, and the evaluative setting to be characterized by more intense emotions compared to the non-evaluative one [16].

3   Method

3.1   Participants

The sample included 83 Italian students (mean age: 25 years, range: 22-56 years; 95% female). They participated to the survey when they were attending the first year of course at the Faculty of Education at the University of Verona, Italy, during the academic year 2011-2012. All the students were contacted again through an on-line methodology during the academic year 2013-2014.

3.2   Material and Procedure

This work is part of a larger project, the "Minimal Knowledge Project", aiming at studying whether and how different cognitive, motivational, and emotional dimensions can predict academic performance in university students attending the Faculty of Education at the University of Verona, Italy. The project started in autumn 2011, and the data has been gathered for three years. In this paper, we present the data about a sub-sample of students who took a compulsory test focused on cognitive abilities they had to demonstrate to possess in order to enrol to their second year of course, including text comprehension abilities among different constructs. The test was administered during their first year of university course, specifically in September, at the beginning of the academic year 2011-2012. The same students were then assessed on motivational and emotional dimensions through an on-line methodology during their third year of course, in July, at the end of the academic year 2013-2014. They also gave the authorization to the researchers to consult their academic career during the entire university period. All the students participated on a voluntary basis and they were assured about anonymity in all the phases of the research. Through this procedure, the three constructs considered in this study have been measured: text comprehension (as a sub-dimension of cognitive abilities), achievement emotions (specifically pride and shame), and academic performance.

3.2.1   Text Comprehension

Text comprehension was one of the constructs included in the "minimal knowledge" test. It was assessed by means of ten questions to be responded choosing among five alternative options. There was only one correct answer (0 = *not correct or missing answer* and 1 = *correct answer*). Therefore, for each participant the total score could range from 0 to 10. The questions referred to the comprehension of two different brief texts. Texts and related questions were randomly extracted from a database used for the "minimal knowledge" test from 2007. The order of the questions was the same for all the students.

3.2.2   Achievement Emotions

Pride and shame were assessed through an instruments developed to evaluate ten achievement emotions, the Brief-Achievement Emotions Questionnaire [see also 1, 15, 17], according to the control-value theory [11, 13]. Achievement emotions were evaluated with an on-line methodology. All the students who had participated to the "minimal knowledge" test were contacted by e-mail and were asked to respond to a questionnaire going to an indicated URL. The students were presented with 30 words (mainly adjectives) describing how they could feel in two different settings–studying and taking exams–and were asked to evaluate how much each word described their feelings, on a 7-point Likert scale (1= *not at all* and 7 = *completely*). These 30 words related to ten achievement emotions, including two positive activity-related emotions (enjoyment, relaxation), three positive outcome-related emotions (hope, pride, relief), two negative activity-related emotions (anger, boredom), and three negative outcome-related emotions (anxiety, shame, hopelessness). The order of the words was randomized and kept constant. The emotions related separately to the two settings–studying and taking exams referred to the university subjects the students were dealing with in that period.

3.2.3 Academic Performance

We obtained from the administrative staff the data related to the total number of passed exams and the grade obtained for such exams (with 18 as the minimum grade and 30 as the maximum grade), until the end of July 2014, for each participant. Therefore, we calculated the mean grade obtained in the exams passed until the end of July 2014.

3.3   Analysis Procedure

We used SPSS version 21.0 for Windows to calculate descriptive statistics, intercorrelations, and analysis of variance (ANOVA). Level of significance was set at $p < .05$. We used Mplus version 5.2 [10] to run path analyses. A maximum-likelihood (ML) estimation was performed, and a bootstrapping method with a confidence interval was used to test indirect effects.

4   Results and Discussion

4.1   Reliability, Descriptive Statistics, and Intercorrelations

For text comprehension test, reliability, estimated with Kuder-Richardson 20 formula, was .77; for achievement emotion, all α-values were higher than .79 (pride - study: α = .90; pride - test: α = .90; shame - study; α = .79; shame - test: α = .85). Such indexes supported homogeneity for each factor.

Table 1. Intercorrelations, means (*M*), and standard deviations (*SD*) about text comprehension, achievement emotions, and academic performance

	1	2	3	4	5	6
1. Text Comprehension	--	--	--	--	--	--
2. Pride - Study	.29**	--	--	--	--	--
3. Pride - Test	.22	.71**	--	--	--	--
4. Shame - Study	-.21	-.28*	-.29**	--	--	--
5. Shame - Test	-.26*	-.33**	-.40**	.62**	--	--
6. Performance	-.27*	.16	.23*	-.21	-.13	--
M	7.33	5.71	6.32	1.76	2.43	25.70
SD	1.59	2.11	1.68	1.25	1.72	1.67

Note. *p < .05, **p < .01.

Students' responses on the ten items about text comprehension were summed, while their responses for the items on pride and shame were averaged together, separately by setting (see Table 1 for descriptive statistics and intercorrelations).

4.2   Path Analysis

We ran a series of path analyses to study the relationships between text comprehension, achievement emotions, and performance, separately for each emotion and for each setting (Figure 1). We included text comprehension as predictor of achievement emotions, in turn predictors of academic performance, to verify the mediating role of the two achievement emotions of pride and shame. Concerning positive emotions, we found that text comprehension positively predicted pride for both settings (study: $\beta =$ .29, $p = .004$; test: $\beta = .22$, $p = .039$), while pride positively predicted academic performance only for test ($\beta = .23$, $p = .029$). As regards negative emotions, text comprehension negatively predicted shame for both settings (study: $\beta = -.21$, $p = .049$; test: $\beta = -.26$, $p = .010$), but shame negatively predicted academic performance only for study ($\beta = -.21$, $p = .047$). For pride related to test, the explained variance of performance was .05; for shame related to study, it was .04 (see Figure 1). To sum up, we found that the emotions of pride and shame played a key role in influencing students' academic performance, with slight differences related to the two settings.

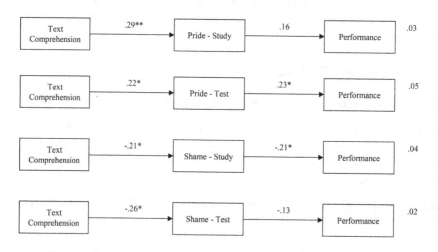

Fig. 1. The standardized paths of the hypothesized models for pride and shame, separately by setting. Explained variances are reported next to performance variables. *$p < .05$, **$p < .01$.

4.3   Analysis of Variance

A repeated-measure analysis of variance (ANOVA), with type of emotion (pride, shame) and setting (study, test) as the within-subject factors, was run, considering the intensity of emotions as the dependent variable. Both a significant effect of emotion, $F(1, 82) = 190.20$, $p < .001$, $\eta_p^2 = .70$, and setting, $F(1, 82) = 34.96$, $p < .001$, $\eta_p^2 = .30$, emerged. On the one hand, intensity was higher for the positive emotion of pride ($M = 6.01$, $SD = 1.75$) compared to the negative emotion of shame ($M = 2.10$, $SD = 1.34$). On the other hand, intensity was lower for the non-evaluative setting, i.e. study ($M = 3.73$, $SD = 1.06$), compared to the evaluative setting, i.e. test ($M = 4.37$, $SD = 0.93$). Such findings support the relevance of examining distinct emotions in different setting to account for learning outcomes.

5   Conclusions

Our findings indicated the relevance of taking into account achievement emotions such as pride and shame, with the advantages associated with the use of an on-line survey, in exploring the relations between a specific cognitive ability like text comprehension and academic performance. Acknowledging limitations related to the nature of self-report measures and composition of our sample, future studies including a wider range of emotions and a larger number of participants, balanced for gender, could examine the generalizability of such effects. From an applied perspective, monitoring continuously emotions through technological instruments such as mobile surveys could help to further understand the complex relationships connecting learning and associated emotions. For example, monitoring how achievement emotions modify within contexts such as Extreme Apprenticeship applied to computer science could be a tool helping to better support learning processes and to sustain task engagement.

References

1. Brondino, M., Raccanello, D., Pasini, M.: Achievement goals as antecedents of achievement emotions: The 3 X 2 achievement goal model as a framework for learning environments design. In: Di Mascio, T., Gennari, R., Vitorini, P., Vicari, R., de la Prieta, F. (eds.) Methodologies and Intelligent Systems for Technology Enhanced Learning. AISC, vol. 292, pp. 53–60. Springer, Heidelberg (2015)
2. Fisher, J.W.: Impacting teachers' and students' spiritual well-being. Journal of Beliefs & Values: Studies in Religion & Education 29(3), 253–261 (2008), doi:10.1080/13617670802465789
3. Goetz, T., Frenzel, C.A., Pekrun, R., Hall, N.C., Lüdtke, O.: Between- and within-domain relations of students' academic emotions. Journal of Educational Psychology 99(4), 715–733 (2007), doi:10.1037/0022-0663.99.4.715
4. Goetz, T., Pekrun, R., Hall, N., Haag, L.: Academic emotions from a social-cognitive perspective: Antecedents and domain specificity of students' affect in the context of Latin instruction. British Journal of Educational Psychology 76(2), 289–308 (2006), doi:10.1348/000709905X42860
5. Graesser, A.C., D'Mello, S.K., Strain, A.C.: Emotions in advanced learning technologies. In: Pekrun, R., Linnenbrick-Garcia, L. (eds.) International Handbook of Emotions in Education, pp. 473–493. Taylor and Francis, New York (2014)
6. Haidt, J.: The moral emotions. In: Davidson, R.J., Scherer, K.R., Goldsmith, H.H. (eds.) Handbook of Affective Sciences, pp. 852–870. Oxford University Press, Oxford (2003)
7. Linnenbrink-Garcia, L., Pekrun, R.: Students' emotions and academic engagement: Introduction to the special issue. Contemporary Educational Psychology 36(1), 1–3 (2011), doi:10.1016/j.cedpsych.2010.11.004

8. Malti, T., Krettenauer, T.: The relation of moral emotion attributions to prosocial and anti-social behavior: A meta-analysis. Child Development 84(2), 397–412 (2013), doi:10.1111/j.1467-8624.2012.01851.x
9. McCallum, F., Price, D.: Well teachers, well students. Journal of Student Wellbeing 4(1), 19–34 (2010)
10. Muthén, L.K., Muthén, B.O.: Mplus user's guide, 5th edn. Muthén & Muthén, Los Angeles (1998–2007)
11. Pekrun, R.: The control-value theory of achievement emotions: Assumptions, corollaries, and implications for educational research and practice. Educational Psychology Review 18(4), 315–341 (2006), doi:10.1007/s10648-006-9029-9
12. Pekrun, R., Goetz, T., Frenzel, A.C., Barchfeld, P., Perry, R.P.: Measuring emotions in students' learning and performance: The Achievement Emotions Questionnaire (AEQ). Contemporary Educational Psychology 36(1), 36–48 (2011), doi:10.1016/j.cedpsych.2010.10.002
13. Pekrun, R., Perry, R.P.: Control-value theory of achievement emotions. In: Pekrun, R., Linnenbrick-Garcia, L. (eds.) International Handbook of Emotions in Education, pp. 120–141. Taylor and Francis, New York (2014)
14. Pekrun, R., Stephens, E.J.: Academic emotions. In: Harris, K.R., Graham, S., Urdan, T., et al. (eds.) APA Educational Psychology Handbook. Individual Differences and Cultural and Contextual Factors, vol. 2, pp. 3–31. American Psychological Association, Washington, DC (2012)
15. Raccanello, D.: Students' expectations about interviewees' and interviewers' achievement emotions in job selection interviews. Journal of Employment Counseling (in press)
16. Raccanello, D., Brondino, M., De Bernardi, B.: Achievement emotions in elementary, middle, and high school: How do students feel about specific contexts in terms of settings and subject-domains? Scandinavian Journal of Psychology 54(6), 477–484 (2013), doi:10.1111/sjop.12079
17. Raccanello, D., Brondino, M., Pasini, M.: Two neglected moral emotions in university settings: Some preliminary data on pride and shame. Journal of Beliefs & Values: Studies in Religion & Education (in press)
18. Raccanello, D., Brondino, M., Pasini, M.: Achievement emotions in technology enhanced learning: Development and validation of self-report instruments in the Italian context. Interaction Design and Architecture (submitted)
19. Tangney, J.P., Stuewig, J., Mashek, D.J.: Moral emotions and moral behavior. Annual Review of Psychology 58, 345–372 (2007), doi:10.1146/annurev.psych.56.091103.070145
20. Tempelaar, D.T., Niculescu, A., Rienties, B., Gijselaers, W.H., Giesbers, B.: How achievement emotions impact students' decisions for online learning, and what precedes those emotions. The Internet and Higher Education 15(3), 161–169 (2012), doi:10.1016/j.iheduc.2011.10.003
21. Tracy, J.L., Robins, R.W.: The psychological structure of pride: A tale of two facets. Personality Processes and Individual Differences 92(3), 506–525 (2007), doi:10.1037/0022-3514.92.3.506

New Experiments on the Nature of the Skillful Coping of a Draftsman

Carolina López[2], Tomás Gómez[1], Manuel Bedia[1], Luis Fernando Castillo[3], and Francisco López[2]

[1] Computer Science Dpt., University of Zaragoza, 50018 Zaragoza, Spain
[2] Plastic and Physical Expression, University of Zaragoza, 50015 Zaragoza, Spain
[3] Engineering Systems Dpt., University of Caldas, Manizales, Colombia
c.lopez@unizar.es

Abstract. During the last years, different authors from disciplines such as cognitive science or psychology have felt interested in analyzing how artists are immersed in an artistic task. This issue is somehow related to the notion of extended cognitive systems. In previous papers, we have showed how it is possible to characterize the "feeling" of a draftsman when he draws through some properties of the hand-tool coupling. Now, we will show new results obtained recently that allow us to define a more complete view on this phenomena and play a crucial role both in the perception either the understanding of the intentions of artists when they are immerse in a sensorimotor process. Finally, we will propose that art teachers can use this knowledge didactically in their classes.

Keywords: art education, motor skills, art experience, psychometric.

1   Introduction

During the last two decades, dynamical systems concepts as self-organized criticality or scale-free patterns have provided new insights about how the brain and the mind operate in a non-linear dynamic manner, self-organizing its activity always at a regime of criticality [2]. These types of patterns have also been widely found in cognitive science and psychology. A particular structure known as 1/f noise represents the intermediate between stability and adaptability and it can be found in a very broad spectrum of scientific fields (for instance, in human gait, brain activity, etc.). Most of them are being useful in clinical applications, particularly for diagnostic purposes since the lack of 1/f patterns seem to be an indicator for disease or malfunctioning in psychology or medicine [12-13]. In this paper we propose to apply fractality indicators to measure and validate aspects related to the phenomena of artistic immersion. The current paper is an extension of previous works where we have applied the same notions and indicators (inspired in [5], where they analyzed the functioning of a mouse during the performance of an experimental sensorimotor task).

The knowledge of the artist's intentions by analyzing the artwork is a classic problem for the history of art. Traditional approaches have been frequently too broad

T. Di Mascio et al. (eds.), *Methodologies & Intelligent Systems for Technology Enhanced Learning,*
Advances in Intelligent Systems and Computing 374, DOI: 10.1007/978-3-319-19632-9_4

because researchers claim to address very general questions but, however, it is unclear that the results could have enough relevance, since it is unlikely that the questions asked could lead to testable hypotheses. However, as result of interdisciplinary frameworks of research in the last years, new formal notions and experimental highlights have been developed. We think they can help us to understand the way and the feeling or the immersion level in which a draftsman works [6-7]. In this paper, we will review very briefly the scientific literature regarding "hand-tool" couplings [4] and, after that, we will refresh the results of previous experiments about how to measure the types of couplings mentioned [1, 11]. Finally, we will propose how to apply those ideas in the domain of drawing and in the art teaching.

2   How to Characterize the Immersion Process of an Artwork

In general, we can define an "interaction-dominant system as a softly assembled system in which any part can take or lose the role of a functional unit of the system, depending upon the richness of physical coupling" [2]. Our proposal (based in a previous work [5]) consists of explaining how long-range correlation can be found in the hand-tool movements and how their values decrease when the situation is temporal restrictive. In this paper we use the same ideas (i) to evaluate the differences in the mode of drawing of several art students using temporal restriction in a sequence of drawing exercises, and (ii) to calibrate the embodied memory in a group of art teachers. As result of our analysis, we will find that the value of the exponent of a $1/f^\beta$ long memory process at the interface between body and tool (a digital pencil that records the trace of a drawing), can be an indicator of different situations of performance in art processes. In [5, 9] the findings come from a simple study of people using a computer mouse designed to malfunction with similar measurements.

3   Experiments

In this paper, following [5], we have used the same set of indicators to measure when a participant is smoothly cope with the pencil while he/she is immerse in a sensorimotor task. In unfavorable situations (in our first experiment, when the participant has got less time to make a drawing; in the second one, when the participants begins a sequence of artworks) the behavior becomes rigid (because the participant focuses the attention on the task and on the drawing). We have proposed two experiments. In the first one, the dynamical structure in the "hand-pencil" movements is analyzed when the temporal restrictions change (tasks that go from 2 min to 5 min). The aim is to measure whether when the participant has more time also shows greater skill measured by values closest to pink fractality. In the second experiment, we observe the evolution of the value of fractal exponents when a group of participants (art teachers) tried to incorporate, in form of a embodied memory, the way of making a drawing. The goal here has been to determine whether the repetition can make fractals processes that characterize sensorimotor tasks more flexible. Both experiments have been developed using the same experimental design.

3.1   Experimental Method

Seventeen undergraduate students from the "School of Arts at Zaragoza" participated in the first experiment and four teachers in the second one. Students ranged in age from 16 to 18 years and teachers from 28 to 32 years. For the experiments, they used Wacom's pens [8], with 2048 levels of pen pressure sensitivity, and the possibility of recording traces and tilt recognition. They allowed us to reproduce with precision the movements of the participant's hands while drawing.

Fig. 1. Picture of the participants in the experiment: Left: an illustration of one of the drawings with a 2 min temporal restriction; Right: illustration with a 5 min temporal restriction

3.2   Measuring the Dependence of a Skillful with Temporal Restrictions

In order to characterize the fluctuations in the movements of the hand-tool systems, we calculate long-range correlations changes which are visualized as $1/f^{\beta}$ type noise in the frequency domain. The significance of $1/f^{\beta}$ noise for behavioral data can be analyzed in [15]. Very briefly, the discussion of $1/f^{\beta}$ noise in behavioral time can be reduced to three typical cases: (i) when the values of the fractal exponent are $0 < \beta < 1$, we have temporal structures near to be random; (ii) if $1 < \beta < 2$, we find out temporal structures revealing very rigid process; (iii) in the case of we have structures when the parameter 'β' approaches to 1, it is an indicative of the activity of a smoothly operating system, softly assembled by virtue of interaction-dominant dynamics.

Procedure, Design and Data Analysis

At the beginning of the experiment, participants were instructed to use the Wacom pencil with both hands. The experimental task consisted of drawing a simple picture (See Figure 1) in two situations: the first one in which the available time was 2 minutes and the second one, when the participants had 5 minutes to finish. The purpose was to analyze whether these temporary differences were reflected in changes of the fractal indicator. During every experimental session, the drawing described by each participants pencil was recorded in a file which also stored a timestamp attribute. From these data (levels of pen pressure sensitivity, recording traces, tilts, etc.) was calculated the fractality exponent. In order to quantify the fractal-like autocorrelation properties of each of the time series generated, a Detrended Fluctuation Analysis (DFA) algorithm was used that estimated the scaling exponent in every fractal series.

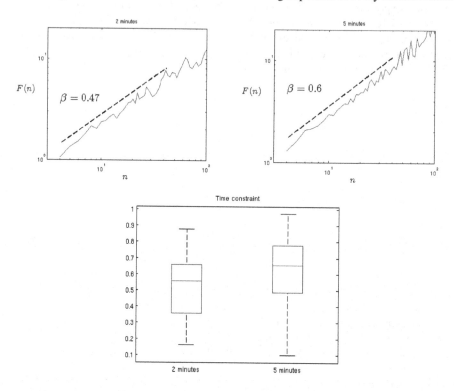

Fig. 2. Upper figures: Fractal exponent β in two cases: a) 2 min temporal restriction, b) 5 min temporal restriction. The examples are representative cases of the two stages in the experiment. Bottom figure: Boxplot distribution of β (population: seventeen students).

The DFA code used in this paper is part of MatLab Toolboxes (code for DFA is available at http://www.nbtwiki.ne). The long-range correlations were analyzed using Detrended Fluctuation Analysis (DFA), a technique which allows us to estimate a coefficient of temporal correlation in a time series [12] that is related to the fractal exponent `β'. A temporal structure formed by input data is divided in different boxes with length N where the sizes change iteratively.

Each temporal structure creates a sequence of fluctuation functions F(n). Inside each block, an analysis of the variability is made taking as a reference a linear regression in a window with size n. In general terms, F(n) increases as n but if we find a case of a linear relationship between them in a log-log plot (as it is shown in Figure 2), then we can assume that a fractal pattern exists in the process and the exponent β in the power-law relation can be obtained through the relationship F(n) vs n (i.e., the scaling parameter can be used to identify persistence of correlated process).

As it is illustrated in Figure 2, both situations show power-law relations but, interestingly, the value of the exponent β is lower in the "2 min temporal restriction case" than in the "5 min case". The results say to us that the degree of skillful when the students have more time to finish is closer to 1 (a symptom of a softly assembled structure, of an interaction-dominant system). As the experiment deals with a group of data that involves common random effects (different drawings) but that, at the same time, represents a level of structure (group of participants) a linear mixed-effect model is used. Briefly, a linear mixed-effect model is a multivariate linear regression model which allow us to take into account the fact that we have different players and repeated measures for each player (they incorporate both fixed and random effects). Mixed effects models were developed using the nlme package in the open-source statistical software R (http://cran.r-project.org/web/packages/nlme/index.html). The mixed model on circular variance values exhibited positively correlated noise diverging from the level characteristic of white noise (p<.001) and separate t-tests indicated that both cases diverged.

3.3   Embodied Memory during an Embodied Artistic Process

Inspired by Varela's ideas [14] the artist Patricia Cain developed in [3] a methodology to investigate the evolution of the practitioner. In [3], she shows that experiential accounts of other drawing practitioners in and interviews provide evidence of a type of "enactive thinking". According to this author, the method of enactive copying allow us to capture something about another's drawing by re-enacting an artwork revealing that what that artist does is not the same of what he was conceptualizing. In [3], the author related that she became consciously aware of what a work was, not from theorizing about what she could see, but from understanding what she physically knew. "Enactive coping" seems to be a way to obtain knowledge from practice, that is, "a way of making a drawing visible what are essentially tacit processes in activity through the ability to make sense of what we do" [3]. Therefore, we are interested in using this methodology in order to check if immersion processes in artwork and sensorimotor tasks can be investigated through this "enactive coping methodology".

Procedure, Design and Data Analysis

In this experiment, 4 teachers (2 females and 2 males) were selected to take part of it. Participants were provided with digital pencils and with instructions for following a drawing plan. They were informed that the experiment consisted in repeating several times the same illustration. We analyzed fractal properties in the work of four art teachers who were asked to draw a picture of Durero (see Figure 3). They repeated the same pattern for several days during a week in order to embodied memorize the way to do it and incorporating memory bodily processes. The experiment comprised a 5-trial training session for each teacher. The variables used in the analysis were the average speed (an indicator of the skill) and average acceleration (an indicator of the intention of the artist). The results, as we will see, indicated a clear increase in the measuring of the fractality. As it is illustrated in Figure 4, both cases (average speed and average acceleration) show power-law relations but the value of the exponent β is closer to 1 when the more practice was made.

Fig. 3. Picture of the participants in the experiment: Illustration of Durero (right) and drawing made for one of the participants (left)

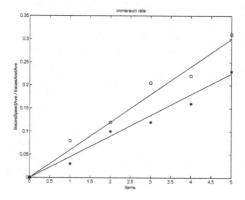

Fig. 4. Evolution of fractal exponent β in an embodied immersion process: a) upper figure: fractality in speed average (variable: tracesSpeedAve) and b) bottom figure: fractality in acceleration average (variable: tracesAcelAve). Both cases are representative of a process of "pinking" of the fractality with a significant increase of the scaling coefficient characterizing long-range correlations (at the same time that more flexible temporal structures).

It is observed an increase of the scaling coefficient (a process of "pinking") characterizing long-range correlations (more flexible temporal structures) when the participants do more a more proofs. Statistical outputs have been obtained using again a mixed model ANOVA.

4   Conclusions

The results of both experiments revealed that the hand-tool coupling, as a part of a larger smoothly interaction-dominant system, becomes functionally better when the students have more time or the teachers have more practice. In both experiments, empirical evidences are found to support the hypothesis that during skilled task performance the behavior of the hand-tool exhibits the kind of power-law scaling $1/f\beta$ associated with the value of the exponent close to one.

When the pencil worked, hand motions followed a mathematical form known as "pink noise" [8]. But when the experimental conditions generated participant's malfunctioned movements (less time, less learning, etc.), the pink noise vanished becoming "white".

In this paper we have proposed that, through metrics able to detect the 'immersion" or "sensitivity" of an artist drawing, it would be possible to get methodological tools capable of measuring aesthetic experiences [16]. And in order to explore our hypothesis, we have analyzed the work of students and teachers in several artistic experiments and thorough the traces of their drawings.

Some empirical evidences are found to support the hypothesis that during skilled task performance the behavior of the hand-tool exhibits the kind of power-law scaling $1/f\beta$ associated with the value of the exponent close to one. Consequently, although one can distinguish anatomically between separate behaving components (i.e., the tool, parts of the body, etc.), the task performance is more appropriately understood by taking the tool to be functionally integrated into an extended body-tool coupling system. It is, in a certain sense, a proof of the thinking is bigger than the biological body limits. We interpret our experimental results within the light of the classical results in cognitive psychology concerning attention and cognitive abilities. We think that our results suggest that the redistribution of attentional resources accompanying a shift from coupling to uncoupling should have an impact on the performance of a demanding cognitive task.

Acknowledgments. This research has been partially supported by the project TIN2011-24660, funded by the Spanish Ministry of science and Innovation, and the project FCT-13-7848, funded by the Spanish Foundation for Science and Technology.

References

1. Albaret, J., Thon, B.: Differential effects of task complexity on contextual interference in a drawing task. Original Research Article Acta Psychologica 100(1-2), 9–24 (1998)
2. Bak, P., Tang, C., Wiesenfeld, K.: Self-organized criticality. Physical Review A 38, 364–374 (1988)
3. Cain, P.: Drawing: The Enactive Evolution of the Practitioner, 1st edn. Intellect (2013) ISBN-10: 1841503258
4. Clark, A.: Being there: Putting brain, body, and world together again. MIT Press, Cambridge (1997)
5. Dotov, D.G., Nie, L., Chemero, A.: A Demonstration of the Transition from Ready-to-Hand to Unready-to-Hand. PLoS ONE 5 (3), e9433 (2010)

6. Dreyfus, H.: Being-in-the-world: A commentary on Heidegger's Being and Time, Division I. The MIT Press, Cambridge (1991)
7. Holden, J., van Orden, G., Turvey, M.: Dispersion of response times reveals cognitive dynamics. Psychological Review 116, 318–342 (2009)
8. Interactive Pen Displays Digital Drawing Tablets, http://www.wacom.com
9. Nie, L., Dotov, D., Chemero, A.: Readiness-to-hand, extended cognition, and multifractality. In: Carlson, L., Hoelscher, C., Shipley, T.F. (eds.) Proceedings of the 33rd Annual Meeting of the Cognitive Science Society, pp. 1835–1840. Cognitive Science Society, Austin (2011)
10. Pinheiro, J., Bates, D.: Mixed-Effects models in S and S-PLUS. Springer, New York (2000)
11. Van Sommers, P.: Drawing and Cognition: Descriptive and Experimental Studies of Graphic Production Processes. Cambridge University Press, Cambridge (1984)
12. Van Orden, G.C., Holden, J.G., Turvey, M.T.: Self-organization of cognitive performance. Journal of Experimental Psychology. General 132(3), 331–350 (2003)
13. Van Orden, G.C., Holden, J.G., Turvey, M.T.: Human cognition and 1/f scaling. Journal of Experimental Psychology: General 134(1), 117–123 (2005)
14. Varela, F.J., Thompson, E., Rosch, E.: The embodied mind: Cognitive science and human experience. MIT, Cambridge (1991)
15. Wilson, R.: Boundaries of the mind: the individual in the fragile sciences: cognition. Cambridge Univ. Press, Cambridge (2004)
16. Wheeler, M.: Reconstructing the cognitive world: The next step. The MIT Press, Cambridge (2005)

Using ICT to Motivate Students in a Heterogeneous Programming Group

Carmen N. Ojeda-Guerra

Department of Telematics Engineering (ULPGC)
carmennieves.ojeda@ulpgc.es

Abstract. ICT has the potential to enhance and transform higher education in many ways and it's seen as important tool to enable and support the move from traditional 'teacher-centric' teaching styles to more 'learner-centric' methods. In this article, we describe and analyze one curricular experiment in the use of ICT in a programming course of an engineering degree. In this course, our students form part of a heterogeneous group with different programming skills. We have checked that the use ICTs in our programming course, increases students' motivation and the examination pass rates.

Keywords: Learning processes and tools, ICT, Problem- and Project-based learning, Motivation.

1 Introduction

Programming is an essential component in the curricula of most technical degrees in universities. Earning a college degree in computer programming can open up a wide range of career opportunities for individuals interested in the ever-growing field of technology [1]. A major in computer programming can prepare students to develop programs that will solve problems, convert data, store and retrieve information, and help individuals communicate via computers. These degree programs teach the student how to bring a computer program into existence, constructing, testing and debugging the program.

Programming requires some degree of aptitude on the part of the students although after nearly four decades of studying, neither researchers nor educators agree on exactly which variables actually predict student success in programming courses [2]. However, one important aptitude is without doubt, the ability to think abstractly, creatively and independently in order to solve general problems using a computer. But, to facilitate learning of new skills, we have to use a teaching/learning methodology, which is based on something common in the diary life of our students. In the past, the structure of the classes involved a master reading from texts and commenting on the readings. The students are silent, passive, and in competition with each other. But, in the current teaching paradigm, with the world moving rapidly into digital media and information, the role of Information and Communication Technologies (ICTs) in education is

© Springer International Publishing Switzerland 2015

T. Di Mascio et al. (eds.), *Methodologies & Intelligent Systems for Technology Enhanced Learning*,
Advances in Intelligent Systems and Computing 374, DOI: 10.1007/978-3-319-19632-9_5

becoming more and more important and this importance will continue to grow and develop in the 21st century.

According to UNESCO [3], ICTs are increasingly utilized by higher education institutions worldwide (for developing course material; delivering and sharing content and so forth). Today's students are digital natives and as such use of technology in education has proven to be effective. Technology infuses classrooms with digital learning tools, such as computers and hand held devices. Effective technology integration is achieved when the use of technology is routine and transparent and when technology supports curricular goals. Technology also changes the way teachers teach [4], offering educators effective ways to reach different types of learners and assess student understanding through multiple means. Many authors agree in develop a specific pedagogy centering on computer programming, such as [5], [6], [7], [8].

On the other hand, assessment procedures in formal education and training have traditionally focused on examining knowledge and facts through formal testing [9], [10], but skills such as problem-solving, creativity, critical thinking, learning to learn, and so forth are becoming increasingly important. In this way, this paper documents a pedagogical approach that is used to improve the outcomes of a computer programming course in an engineering degree (non-computer science). In this course, we use Java language (an Object-Oriented Language) because OO programming has become a main paradigm of programming in industry, as well as in computer science and engineering education. As reported by [11], more than 85% of all introductory programming courses include object-oriented concepts, and of these more than 60% are objects-first courses. Also, we integrate the technology into our lessons and the students' assignments (course projects, which are proposed by the students). Our main goal is to motivate our students to learn how to implement medium complexity programs in Java language, and achieve that those students lose the fear to solve real-world programming problems.

The rest of the paper is organized as follows. In next section, we outline some important methodologies of learning. Next, we introduce the proposed methodology and briefly describe the process that we follow to motivate our students in their learning. Next, we present how we assess this process. Finally, we conclude by outlining the main goals of our approach.

2   Models of Learning

Introductory programming subjects traditionally have high failure rates and as they tend to be core to IT and Computer Science courses can be a road block for many students to their university studies. The result is that attrition rates are high (estimated by [12] to be between 30% and 60%). Due to this, computer programming teachers need models of learning which decrease these rates.

According to [13], teaching and learning are forever intertwined (a single process, as seen from two different vantage points). The students' learning support can be realized through pedagogical approaches, such as: *Learning by Doing* (students must practice in order to familiarize themselves with programming. Starting with the main concepts given by teachers, students can learn new skills), *Learning by Teaching* (the best way to learn is by teaching others. It's an effective way to reinforce a student's knowledge), and *Independent learning* (shift of responsibility for the learning process from the teacher to the student. The learner is autonomous, and separated form his/her teacher by space and time. In this approach, students and teachers must have some method of communication).

In that context, the problem-based learning (PBL) and problem- and project-based learning (PPBL) approach are examples of learning by doing and depending on the teacher's role, they can be examples of independent learning, too. In these approaches, educators take on the role of facilitator, guiding the students learning and monitoring their progress in different ways. Many authors as [14], [15], [16] use these approaches in different disciplines of science.

However, whatever the approach or methodology used in a course, the use of technology cannot be avoided at all. The most important benefit from ICT in education it is hoped for is improved learning outcomes and decreased the attrition rate. In this context, there are many examples available online about the use of ICT in many universities around the world. The common theme in most of them is that teachers are expected to know to successfully integrate ICT into his/her subject areas to make learning more meaningful.

3   Proposed Model of Learning

Most existing computer programming instruction mainly focuses on the training of programming language syntax and programming skills, while the problem-solving concepts are often ignored [17]. In Programming, students should be advised to maximize their time by trying to solve problems and absorb concepts on their own [13].

On the other hand, the growing use of ICT as an instructional medium is changing many of the strategies employed by both teachers and students in the learning process. In this article, we proposed a model of learning based on ICT in order to increase the learning motivation and improve the programming skills which leads to decrease of the attrition rates. Our model of learning is used in a programming course in an engineering degree (non-computer science degree) with good results. In this degree, there are only 3 four-months subjects (60 hours per subject) about programming: two of them about basic programming (in the first year) and one about advanced programming (in the third year). In the basic programming courses, the main problem for teachers is the low motivation of the students because they think that those skills are not necessary in their future work, and also, some of them enroll in the degree only as a last resort (for teachers in these subjects, it's very difficult to reach good results).

On this basis, in the third year of the degree (with a long blank year in the middle), we have the following problems: a) Students who have failed the assessment in the basic subjects, so they are enrolled in the basic and advanced programming courses (around 15%); b) Students who have passed the assessment but they don't remember many programming concepts (around 76%) and c) Students who have passed the assessment and have programming skills (around 9%). In addition to the above, in the third year of the degree, our students have to choose their engineering qualification (in the degree, we have four different qualifications), and only one of them is driven towards programming computers and devices, so the programming course in the third year is the last programming course that around 75% of our students will take in the degree.

Against this gloomy background, we have developed a model of learning which seeks to take into account the specific characteristics of our heterogeneous group of students. Our model is an example of problem- and project-based learning (PPBL) and consists of two different learning models: learning by doing (problem-based approach) and independent learning (project-based approach). In this methodology, we take the three first weeks to "warm up" students, providing different exercises to remind them the basic OOP concepts and keeping students in their comfort level. After this, we use the learning by doing and independent learning approaches.

3.1   Learning by Doing

In this point, our purpose is to teach (or remind) the main concepts in advanced programming using two approaches:

- By non-lab teaching (around 20 hours): we present the information using animated slides that can be later used by students to construct their own knowledge, so they must gives students a good idea about the topic, independent from the lecture. Also, we use the blackboard to emphasize some concepts or implement some example codes. Our students use their laptops or tablets to take notes and implement the codes at the same time.
- By lab teaching (around 30 hours): in this approach, we make two different type of exercises: a) hard guided and b) soft guided. In the first case, we implement several programming exercises using a computer in front of the students. We use a projector and students learn new concepts by observing the way in which the teacher solves the programming problem, using a computer and an IDE. In the second case, we give our students a short textual description of the expected behavior of the programs (finding an adequate balance for the amount of guidance given to students is one of the challenges of this type of educational approach). Teachers are facilitators of learning, guiding the learning process and promoting an environment of inquiry. Students use step-by-step problem-solving skills, and at the same time, they are pulling information from their own memory, experience, and base knowledge to master the new programs. This trains the routine and gives a constant feeling of success by achieving small goals.

In the lab teaching approach, the major problem was that our classrooms are big and we have many students in them so that beyond the second row, anybody can see the screen. In order to help students to follow lessons, we designed a new tool called *Binocular* tool that is an Eclipse plugin. This tool doesn't work equally in the teacher's computer than in the students' computers. Students open a new Eclipse view called *Binocular view* (similar to *Console view*) that will show the program that is being implemented by the teacher (each student can configure the text size). On the other hand, the teacher runs the tool and configures the refresh time. Any change in the teacher's code, it will appear in the students' view (although students can see the teacher's code in their own view, they can not copy and paste this code).

3.2   Independent Learning

In this step, the main challenge is to increase the learning motivation of our students without forgetting to reach the learning goals that enable people to attain new information and more complex knowledge. Numerous studies have shown that the motivation has a remarkable effect on learning [18]. To achieve this motivation, we have to find a suitable set of integrative activities (projects) for students, which use most important programming concepts. These projects must be exciting but, at the same time, useful. In order to do that, we give a specification of the required job to our students and they have to propose us a new project (one per student) that matches with our requirements. In the specification, we focus on the concepts explained in the exercises implemented in the classroom, but we encourage our students to go far beyond the specification.

We request two different projects along the course. Each project must be implemented in 4 weeks and must be realistic. An example of second final project can be: *"Designing a web application that accesses the transceivers information (the transmitted and received power, location, manufacturer, technology ...) of a company x"*. The student has to design and implement: the database (using MySQL), the classes to interact with the database (using JDBC and Java), the servlets (using Netbeans or Eclipse and Java) and the web pages (using HTML). All the concepts involve in the final project were explained in the learning by doing approach. Before starting each project, students have to submit a report in which they explain what they want to do. For example, for the second project, students have to explain: tables and fields of the database, classes to interact with the database and their members (attribute and methods) and format of the different web pages (at least there should be two).

Also, at the end of each project, students have to submit the results. In the first project, they have to submit the implemented code properly commented. For the second project, students must produce a short tutorial video (5-8 minutes long) in which, the student explains: the design, implementation and testing process. Doing this video, students learn to: Use multimedia technology (not only as a user but as a producer) which helps them harness their creativity; design/plan a video "movie", in which the student is the presenter. This plan must show, in a right order, the processes described above; synthesize the information; add subtitles, zoom, shortcuts, transitions... Anything that can make the viewing easier and enjoyable and announce objectives, show results and explain possible mistakes.

4   Assessment Procedures

Assessment procedures in formal education and training have traditionally focused on examining knowledge and facts through formal testing [10] (similar to intelligent tests) and do not easily lend themselves to grasping "soft skills". This traditional procedures need to be revised to more adequately reflect the skills needed for life in the 21st century. In this context, we use two different strategies:

- Inside classroom: our assessments consist of two incomplete Java codes. In different parts of these codes, we include TODO comments specifying what students have to do (figure 1). If they update the codes correctly, the codes will pass the provided JUnit tests. In these assessments, students can use any documentations and can access to Internet because in real life, if any programmer wants to solve a problem, he/she accesses to Internet and specialized documentation if it's needed (increasing the looking for documentation skills).

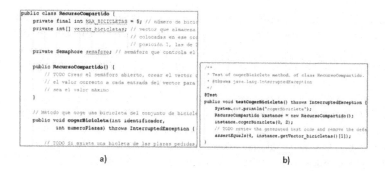

Fig. 1. Example of assessment inside classroom: a) code and b) test

In order to avoid the plagiarism, we take the following decisions: we test around 20 students at the same time and don't let them open the email service (for inside classroom assessments) and don't allow two students have the same final projects (for outside classroom assessments). However, in the last case, we allow the students cooperate each other.
- Outside classroom: we use the initial and final reports of each project submitted by students in order to compare the initial specification with the final result. Also, the submitted video must capture the audience attention and give them a good learning experience.

In figures 2(a) and 2(b), we present the final results of this year in the basic and advanced programming courses respectively (in the previous years, the results are very similar). The field "Others" in the first one corresponds with those students who dropped out the degree and in the second one corresponds with the students who dropped out the course in that year. We can see that with

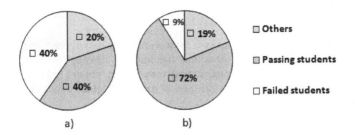

Fig. 2. Final result in: a) the basic courses and b) the advanced course

the same kind of students, we obtain much better results in the advanced course than in the basic course because our students are more motivated. In our course, more than 70% of students finalizes the course with success. Also, we have made a study about why some students drop out the course, and the conclusions say us that all of these students are in the 15%, who have failed the assessment in the basic subjects.

Most students are very motivated and, some of them who started with gaps in skills, implement very complex projects. However, there are still students that present low quality final jobs. In our opinion, this is due to they are not interested in computer programming (the course is compulsory and they have to take it) or in the worst case scenario, they can't learn to program [2].

5 Conclusion

In this article, we describe and analyze one curricular experiment in the use of ICT in a programming course of an engineering degree (non-computer science with only 3 different four-months subjects about programming). In this course, our students form part of a heterogeneous group with different programming skills and in many cases, with low motivation.

On the other hand, although Programming should be taught rigorously it's important and necessary to take some decisions in order to decrease the dropout rate and increase the success of the outcomes, encouraging learners to take control of their learning. To do this, we use a model of learning (that is an example of problem- and project-based learning or PPBL) which consists of two different learning models: learning by doing (problem-based approach) and independent learning (project-based approach). By problem-based learning, we reinforce the students' basic knowledge in programming and by project-based learning we encourage students to research in the real-world problem, that is very positive and gives them extra confidence in their skills.

Also, we have proven that using our methodology, we reach good final results in comparison with the results obtained in the basic programming courses of our degree, taking into account that: at the beginning of the course, most of our students don't remember the basic concepts and there is a blank year (a long gap) between the basic courses and our course.

References

1. Web site of WorldWideLearn, `http://www.worldwidelearn.com/`
2. Dehnadi, S., Bornat, R.: The camel has two humps (2006),
   `http://www.cs.mdx.ac.uk/research/PhDArea/saeed/paper1.pdf`
3. Web site of UNESCO, `http://www.unesco.org/en/higher-education/`
   `higher-education-and-icts/`
4. How Technology is Changing the Ways Students Learn and Teachers Teach. White
   paper by Motorola (2010)
5. Chetty, J., van der Westhuizen, D.: Toward a Pedagogy Centering on Computer
   Programming for Learners in South Africa: An Educational Design Research Approach. In: Proc. of the EdMedia 2014 (2014)
6. Goteti, P., Vignan, S.: A Programming Pedagogy for First Level Programming
   Course using Requirement Driven Approach. In: Proc. of the 3rd International
   Conference on Education and New Learning Technologies (2011)
7. Zhang, X., Zhang, C., et al.: Teaching Introductory Programming to IS Students:
   The Impact of Teaching Approaches on Learning Performance. Journal of Information Systems Education 24(2), 147–155 (2013)
8. Moura, I., van Hattum-Janssenb, N.: Teaching a CS introductory course: an active
   approach. Journal Computers & Education archive 56(2), 475–483 (2011)
9. Redecker, C., Johannessen, O.: Changing Assessment — Towards a New Assessment Paradigm Using ICT. European Journal of Education 48(1) (2013)
10. Cachia, R., Ferrari, A., et al.: Creative Learning and Innovative Teaching: Final
    Report on the Study on Creativity and Innovation in Education in EU Member
    States (Seville, JRC-IPTS) (2010)
11. Bennedsen, J., Schulte, C.: A competence model for object-interaction in introduc-
    tory programming. In: Proc. of the 18th Workshop of the Psychology of Program-
    ming Interest Group (2006)
12. Robins, A.: Learning edge momentum: A new account of outcomes in CS1. Com-
    puter Science Education 20(1), 37–71 (2010)
13. Zhao, G.: On Effective Pedagogical Practice in Teaching Computer Programming
    to CS/CIS Majors. In: Proc. for the Northeast Region Decision Sciences Institute
    (April 2013)
14. Wiek, A., Xiong, A., et al.: Integrating problem- and project-based learning into
    sustainability programs: A case study on the School of Sustainability at Arizona
    State University. International Journal of Sustainability in Higher Education 15(4),
    431–449 (2014)
15. Yadav, A., Subedi, D., et al.: Problem-based Learning: Influence on Students'
    Learning in an Electrical Engineering Course. Problem-based Learning: Influence
    on Students' Learning in an Electrical Engineering Course 100(2), 253–280 (2011)
16. O'Grady, M.: Practical Problem-Based Learning in Computing Education. Journal
    ACM Transactions on Computing Education 12(3) (2012)
17. Wang, H., Huang, I., Hwang, G.: Effects of an Integrated Scratch and Project-
    Based Learning Approach on the Learning Achievements of Gifted Students in
    Computer Courses. In: Proc. of the 3rd International Conference on Advanced
    Applied Informatics (2014)
18. Bergin, S., Reilly, R.: The influence of motivation and comfort-level on learning to
    program. In: Proc. of PPIG 2005 (2005)

Quality of Learning under an All-Inclusive Approach

José Neves[1,*], Margarida Figueiredo[2], Lídia Vicente[3], Guida Gomes[1],
Joaquim Macedo[1], and Henrique Vicente[4]

[1] Algoritmi, Universidade do Minho, Braga, Portugal
{jneves,macedo}@di.uminho.pt, mguida.mgomes@gmail.com
[2] Departamento de Química, Centro de Investigação em Educação e Psicologia,
Escola de Ciências e Tecnologia, Universidade de Évora, Évora, Portugal
mtf@uevora.pt
[3] Agrupamento de Escolas de Reguengos de Monsaraz, Reguengos de Monsaraz, Portugal
lmrcvicente@gmail.com
[4] Departamento de Química, Centro de Química de Évora, Escola de Ciências e Tecnologia,
Universidade de Évora, Évora, Portugal
hvicente@uevora.pt

Abstract. Learning, knowledge, educations are syntax forms that stand for a multifaceted matter, and its assets set the advances on culture, organization, and social matters of any society. However, it is not enough just to instruct, it is necessary to do it with quality, in a holistic way, in order to develop academic and social skills. From this point of view, the weight of the formal, non-formal and informal learning contexts should be underlined, and the use of defective information must be emphasizing. Under this setting the assessment to quality of learning is mandatory, although it is hard to do with traditional methodologies for problem solving. Indeed, in this work we will focus on the development of a decision support system, in terms of its knowledge representation and reasoning procedures, under a formal framework based on Logic Programming, complemented with an approach to computing centered on Artificial Neural Networks, to evaluate the Quality of Learning and the respective Degree-of-Confidence that one has on such a happening.

Keywords: Learning, Knowledge, Education, Logic Programming, Knowledge Representation and Reasoning, Artificial Neuronal Networks.

1   Introduction

The educational directives defined by the European Union at the beginning of the 21st century highlight the importance of education and training processes to assure European competitiveness in an open and globalized world [1]. To unlock the wider benefits of education, all children need the chance to complete not only primary school but also the lower secondary one. And access to schooling is not enough on its own; indeed, education needs to be of good quality so that children actually learn. Given

---

* Corresponding author.

T. Di Mascio et al. (eds.), *Methodologies & Intelligent Systems for Technology Enhanced Learning,*
Advances in Intelligent Systems and Computing 374, DOI: 10.1007/978-3-319-19632-9_6

education's transformative power, it needs to be a central part of any post-2015 global development agenda [2].

The investment on a quality education system at all levels is therefore an imperative in the current society. In order to prepare our students we have to go beyond the basics both in terms of academic skills and social abilities [3]. To achieve this goal, education must be seen in a holistic way, considering the learning integrated into the curriculum that students do in the school (formal education), out of school (non-formal education), and on extra-curricular learning, that students make in interaction with others in the community (informal education).

In a formal education context all the knowledge acquired in the different disciplines and all the skills developed through carrying out the various activities proposed to the student by the teacher are important. Additionally, the relationships in school, i.e., the interactions students – students; students – adults; adults – adults, influences the success of students and teachers [4].

In a context of non-formal education students can achieve, out of school, learning projects that complement and consolidate the learning done in the classroom. Study visits, conducting outdoor projects, small investigations made in any outdoor space to school, should be considered. These extracurricular activities promote the development of skills like adaptability, originality, creativity, promptness, spontaneity, and a quickness in the intellectual processes [5].

Finally, informal education refers to all other learning and skills acquired. Comprises the learning made through the media, in family, and through relationships with others in the community.

In the current European context, education is seen as a key to sustainable development. Growth and competitiveness debates highlight the central role given to education and educational system, formal, non-formal and informal learning to ensure a sustained path of economic development and social cohesion [6].

The stated above shows that the *Quality of Learning* (*QoL*) should be correlated with all actors involved in all situations of the educational process, namely the students, the family, the school and the community. Consequently, it is difficult to assess to the *QoL* since it needs to consider different conditions with intricate relations among them, where the available data may be incomplete, contradictory and/or unknown. In order to overcome these drawbacks, the present work reports the founding of a computational framework that uses knowledge representation and reasoning techniques to set the structure of the information and the associate inference mechanisms. We will centre on a *Logic Programming* (*LP*) based approach to knowledge representation and reasoning [7, 8], complemented with a computational framework based on *Artificial Neural Networks* (*ANNs*) [9].

2   Knowledge Representation and Reasoning

Many approaches to knowledge representation and reasoning have been proposed using the Logic Programming (LP) paradigm, namely in the area of Model Theory [10, 11], and Proof Theory [7, 8]. In this work it is followed the proof theoretical approach in terms of an extension to the LP language. An Extended Logic Program is a finite set of clauses in the form:

$$p \leftarrow p_1, \cdots, p_n, not \ q_1, \cdots, not \ q_m \qquad (1)$$

$$? \ (p_1, \cdots, p_n, not \ q_1, \cdots, not \ q_m) \ (n, m \geq 0) \qquad (2)$$

where "?" is a domain atom denoting falsity, the p_i, q_j, and p are classical ground literals, i.e., either positive atoms or atoms preceded by the classical negation sign $\neg$ [7]. Under this emblematic formalism, every program is associated with a set of abducibles [10, 11] given here in the form of exceptions to the extensions of the predicates that make the program.

Due to the growing need to offer user support in decision making processes some studies have been presented [12, 13] related to the qualitative models and qualitative reasoning in Database Theory and in Artificial Intelligence research. With respect to the problem of knowledge representation and reasoning in LP, a measure of the *Quality-of-Information* (*QoI*) of such programs has been object of some work with promising results [14, 15]. The *QoI* with respect to the extension of a predicate i will be given by a truth-value in the interval [0, 1].

On the one hand, it is now possible to engender the universe of discourse, according to the information given in the logic programs that endorse the information about the problem under consideration, according to productions of the type:

$$predicate_i - \bigcup_{1 \leq j \leq m} clause_j(x_1, \cdots, x_n) :: QoI_i :: DoC_i \qquad (3)$$

where U and m stand, respectively, for *set union* and the *cardinality* of the extension of *predicate_i*. On the other hand, DoC_i denotes one's confidence on the attribute`s values of a particular term of the extension of *predicate_i*, whose evaluation is given in [16].

3   A Case Study

In order to exemplify the applicability of our problem solving methodology, we will look at the relational database model, since it provides a basic framework that fits into our expectations, and is understood as the genesis of the *LP* approach to Knowledge Representation and Reasoning [7].

As a case study, consider the scenario where a relational database is given in terms of the extensions of the relations depicted in Fig. 1, which stands for a situation where one has to manage information about the *QoL*. Under this scenario some incomplete and/or unknown data is also available. For instance, in *General Characterization* database the *Household incomes* for case 3 is unknown, while for example 1 ranges in the interval [2800, 3200].

			Student Related Factors			
#	Absenteeism	Indiscipline	Victim of Bullying	Lack of school ambitions	Alcohol Consumption	Drugs Consumption
1	0	0	0	0	0	0
...	...	...	...	...	...	...
n	1	1	0	1	0	0

				General Information		
#	Age	Gender	Grade	Course	Household members	Household Income (€)
1	15	F	9°	Basic	4	[2800,3200]
...	...	...	...		...	...
n	17	M	11°	Secondary	5	⊥

			Quality of Learning (QoL)				
#	Age	Gender	Income per Capita	Student Issues	School Issues	Community Issues	Family Issues
1	15	0	[700,800]	0	4	5	⊥
...	...	...	...	...	...	...	...
n	17	1	⊥	3	2	0	8

			School Related Factors				
#	Alternative curricula	Programs to support students with difficulties	Reduced number of students per class	Psychological support to students	Non-overloaded schedules	Strong binding with family
1	1	1	0	1	0	1
...	...	...	...	...	...	...
n	1	0	0	1	0	0

			Community Related Factors			
#	Outside school activities	Cultural activities	Offer of private lessons	Strong school-community interaction	Good transport accessibility	High qualifications requirement
1	1	1	1	0	1	1
...	...	...	...	...	...	...
n	0	0	0	0	0	0

	Family Related Factors						
#	Schooling		Socio-economic/cultural status	Parental involvement in school	Parental involvement in extra-scholar activities	Absence of alcohol	Absence of Drugs
	Father	Mother					
1	3	2	2	⊥	⊥	1	1
...	...	...	...	...	...	...	...
n	1	2	1	1	1	1	1

Fig. 1. An Extension of the Relational Database model. In column *Gender* of *QoL* database, 0 (zero) and 1 (one) stand, respectively, for *female* and *male*. In the last two columns of *Family Related Factors* database 0 (zero) denotes *no* and 1 (one) denotes *yes*.

The *Student, School* and *Community Related Factors* databases are populated with 0 (zero) and 1 (one), denoting *absence/no* and *presence/yes*, respectively. In *Family Related Factors* database the column *Schooling* ranges in the interval [0, 3], wherein 0 (zero), 1 (one), 2 (two) and 3 (three) denote, respectively, *illiterate, primary education, secondary education* and *university course*. The values presented in *Socio-Economic/Cultural Status, Parental Involvement in School* and *Parental Involvement in Extra-Scholar Activities* columns range in the interval [0, 2], where 0 (zero), 1 (one) and 2 (two) stand for *Low, Medium* and *High*, respectively.

The values present in the *Student, School, Community* and *Family Issues* in *QoL* database are the sum of the respective values, ranging between [0, 6] for the three first cases and between [0, 14] for the last one. Now, we may consider the relations given in Fig. 1, in terms of the *qol* predicate, depicted in the form:

$$qol: Age, G_{ender}, Inc_{ome\ per\ capita}, Stud_{ent\ Issues}, Sch_{ool\ Issues},$$

$$Com_{munity\ Issues}, Fam_{ily\ Issues} \rightarrow \{0,1\}$$

where *qol* stands for the predicate *quality of learning*, where 0 (zero) and 1 (one) denote, respectively, the truth values *false* and *true*. Thus, from the extension of the quality of learning relation (Fig. 1) it is possible to apply the normalization algorithm presented in [16] to the information regarding case 1 (one), where the former clause denotes the closure of predicate *qol*:

Begin,

The predicate's extensions that make the Universe-of-Discourse are set $\leftarrow$

{

$\neg qol\ (Age, G, Inc, Stud, Sch, Com, Fam)$

$\leftarrow not\ qol\ (Age, G, Inc, Stud, Sch, Com, Fam)$

$$qol\left(\underbrace{15,\quad 0,\quad [700,800],\quad 0,\quad 4,\quad 5,\quad \bot}_{attribute's\ values}\right) :: 1 :: DoC$$

$$\underbrace{[6,22][0,1][200,3000][0,6][0,6][0,6][0,14]}_{attribute's\ domains}$$

} :: 1

The attribute's values ranges are rewritten ←

{

$\neg qol\ (Age, G, Inc, Stud, Sch, Com, Fam)$

$\qquad\qquad\qquad\qquad\qquad$ ← $not\ qol\ (Age, G, Inc, Stud, Sch, Com, Fam)$

$qol\left(\underbrace{[15,15], [0,0], [700,800], [0,0], [4,4], [5,5], [0,14]}_{attribute's\ values}\right)$ :: 1 :: DoC

$\qquad\quad\underbrace{[6,22]\ \ [0,1][200,3000]\ [0,6]\ [0,6]\ \ [0,6]\ \ [0,14]}_{attribute's\ domains}$

} :: 1

The attribute's boundaries are set to the interval [0,1] ←

{

$\neg qol\ (Age, G, Inc, Stud, Sch, Com, Fam)$

$\qquad\qquad\qquad\qquad\qquad$ ← $not\ qol\ (Age, G, Inc, Stud, Sch, Com, Fam)$

$qol\left(\underbrace{[0.56,0.56], [0,0], [0.18,0.21], [0,0], [0.67,0.67], [0.83,0.83], [0,1]}_{attribute's\ values\ ranges\ once\ normalized}\right)$ :: 1 :: DoC

$\qquad\underbrace{[0,1]\quad [0,1]\quad [0,1]\quad [0,1]\quad [0,1]\qquad [0,1]\quad [0,1]}_{attribute's\ domains\ once\ normalized}$

} :: 1

The DoC's values are evaluated ←

{

$\quad\neg qol\ (Age, G, Inc, Stud, Sch, Com, Fam)$

$\qquad\qquad\qquad\qquad\qquad$ ← $not\ qual\ (Age, G, Inc, Stud, Sch, Com, Fam)$

$qol\left(\underbrace{1,\qquad 1,\qquad 0.999,\quad 1,\qquad 1,\qquad\quad 1,\qquad\quad 0}_{attribute's\ confidence\ values}\right)$ :: 1 :: 0.86

$\quad\underbrace{[0.56,0.56], [0,0], [0.18,0.21], [0,0], [0.67,0.67], [0.83,0.83], [0,1]}_{attribute's\ values\ ranges\ once\ normalized}$

$\qquad\underbrace{[0,1]\quad [0,1]\quad [0,1]\quad [0,1]\quad [0,1]\qquad [0,1]\quad [0,1]}_{attribute's\ domains\ once\ normalized}$

} :: 1

End.

where its terms make the training and test sets of the *Artificial Neural Network* (*ANN*) given in Fig. 2.

4   Artificial Neural Networks

Several studies have shown how *Artificial Neural Networks* (*ANNs*) could be success-fully used to structure data and capture complex relationships between inputs and outputs [17, 18]. *ANNs* simulate the structure of the human brain being populated by multiple layers of neurons. As an example, let us consider the former case presented in Fig. 1, where one may have a situation in which the assessment to learning quality is needed. In Fig. 2 it is shown how the normalized values of the interval boundaries and their *DoC* and *QoI* values work as inputs to the *ANN*. The output translates the *QoL* and the confidence that one has on such a happening. In addition, it also contrib-utes to build a database of study cases that may be used to train and test the *ANN*.

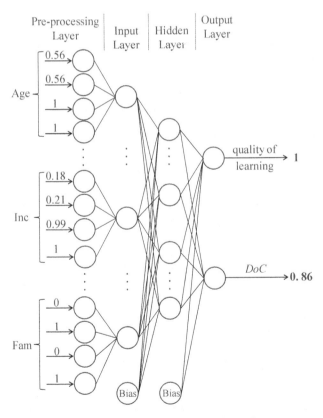

Fig. 2. The Artificial Neural Network topology

The dataset holds information about the factors considered critical in the prediction of *QoL*. Thirteen variables were selected allowing one to have a multivariable dataset with 548 records. These variables were grouped into five main categories, i.e., *General Information* and *Student, School, Family* and *Community Issues* (Fig. 1). Thus, the number of variables used as input of the ANN model was reduced to seven (Table 1). The gender distribution was 51.4% and 48.6% for female and male, respectively. To ensure statistical significance of the attained results, 20 (twenty) experiments were applied in all tests. In each simulation, the available data was randomly divided into two mutually exclusive partitions, i.e., the training set with 67% of the available data, used during the modeling phase, and the test set with the remaining 33% of the cases, used after training in order to evaluate the model performance and validate it. The back propagation algorithm was applied in the learning process of the *ANN*. The activation function used in the pre-processing layer was the identity one. In the other layers was used the sigmoid activation function.

A common tool to evaluate the results presented by the classification models is the coincidence matrix, a matrix of size $L \times L$, where L denotes the number of possible classes. This matrix is created by matching the predicted and target values. L was set to 2 (two) in the present case. Table 2 present the coincidence matrix (the values denote the average of the 20 experiments). A perusal of Table 2 shows that the model accuracy was 95.4% for the training set (352 correctly classified in 369) and 94.4% for test set (169 correctly classified in 179).

Table 1. Variables description

Variable	Description
Age	Student's age
Gender	Student's gender
Income per capita	Monthly income per capita of the household in euro
Student Issues	Includes the issues related with the student himself and can influence the *QoL*, like absenteeism, indiscipline, bullying, lack of school ambitions, alcohol and/or drugs consumption
School Issues	Comprises the topics linked with the school and can affect the *QoL*, such as the existence of alternative curricula, programs to support students with difficulties, reduced number of students per class, psychological support to students, non-overloaded schedules or strong binding with family
Community Issues	Take into account the community related factors that can determine the *QoL*, like the outside school activities, cultural activities, offer of private lessons, strong school-community interaction, good transport accessibility or high qualifications requirement
Family Issues	Contains familiar characteristics that can influence the *QoL*, such as parents' schooling, socio-economic and cultural status, parental involvement in school, parental involvement in extra-scholar activities, parental alcohol and/or drugs consumption

Table 2. The coincidence matrix for ANN model

Target	Predictive			
	Training set		Test set	
	False (0)	True (1)	False (0)	True (1)
False (0)	103	12	49	7
True (1)	5	249	3	120

5   Conclusions and Future Work

A *QoL* performance measurement and improvement model would not only be an inestimable process or practice, but something that has to have its roots or become part of the in-structure of education. When the educational practice is intertwined with its simultaneous evaluation as to its impact on the students' learning, then we may have a true discussion about quality, accounting and accountability of education.

Indeed, on the one hand, the involvement of all actors, in all the circumstances of the educational process, aiming the constant improvement the *QoL*, are critical factors associated with successful programs and are an added value to the schools Educational Projects. On the other hand, once the parameters to assess *QoL* are not fully represented by objective data (i.e., are of types unknown or not permitted, taken from a set or even from an interval), the problem was put into the area of problems that must be tackled by Artificial Intelligence based methodologies and techniques for problem solving. In fact, the computational framework presented above uses powerful knowledge representation and reasoning methods to set the structure of the information and the associate inference mechanisms. One's approach may revolutionize prediction tools in all its variants, making it more complete than the existing ones. It enables the use of normalized values of the interval boundaries and their respective *QoI* and *DoC* values, as input to the *ANN*. The output translates the *QoL* and the confidence that one has on such a happening.

Future work may recommend that the same problem must be approached using others computational frameworks like Case Based Reasoning [19], Genetic Programming [8] or Particle Swarm [20], just to name a few.

Acknowledgments. This work has been supported by FCT – Fundação para a Ciência e Tecnologia within the Project Scope UID/CEC/00319/2013.

References

1. Dias, M.: The Impact of Lisbon's Strategy on the Patterns of Education and Training in Portugal. Procedia – Social and Behavioral Sciences 116, 1885–1889 (2014)
2. United Nations Educational, Scientific and Cultural Organization: Making Education a Priority in the Post-2015 Development Agenda. UNESCO Edition (2013)
3. Fitzgerald, C.J., Laurian, S.: Caring our Way to more Effective Learning. Procedia – Social and Behavioral Sciences 76, 341–345 (2013)

4. Gordon, R., Preble, B.: Transforming school climate and learning: Beyond bullying and compliance. Corwin Press, Thousand Oaks (2011)
5. Gloria, R., Tatiana, D., Constantin, R.B., Marinela, R.: The Effectiveness of Non-formal Education in Improving the Effort Capacity in Middle-school Pupils. Procedia – Social and Behavioral Sciences 116, 2722–2726 (2014)
6. Tudor, S.L.: Formal – Non-formal – Informal in Education. Procedia – Social and Behavioral Sciences 76, 821–826 (2013)
7. Neves, J.: A logic interpreter to handle time and negation in logic data bases. In: Muller, R.L., Pottmyer, J.J. (eds.) Proceedings of the 1984 Annual Conference of the ACM on the Fifth Generation Challenge, pp. 50–54. ACM, New York (1984)
8. Neves, J., Machado, J., Analide, C., Abelha, A., Brito, L.: The halt condition in genetic programming. In: Neves, J., Santos, M.F., Machado, J.M. (eds.) EPIA 2007. LNCS (LNAI), vol. 4874, pp. 160–169. Springer, Heidelberg (2007)
9. Cortez, P., Rocha, M., Neves, J.: Evolving Time Series Forecasting ARMA Models. Journal of Heuristics 10, 415–429 (2004)
10. Kakas, A., Kowalski, R., Toni, F.: The role of abduction in logic programming. In: Gabbay, D., Hogger, C., Robinson, I. (eds.) Handbook of Logic in Artificial Intelligence and Logic Programming, vol. 5, pp. 235–324. Oxford University Press, Oxford (1998)
11. Pereira, L.M., Anh, H.T.: Evolution prospection. In: Nakamatsu, K., Phillips-Wren, G., Jain, L.C., Howlett, R.J. (eds.) New Advances in Intelligent Decision Technologies. SCI, vol. 199, pp. 51–63. Springer, Heidelberg (2009)
12. Halpern, J.: Reasoning about uncertainty. MIT Press, Massachusetts (2005)
13. Kovalerchuck, B., Resconi, G.: Agent-based uncertainty logic network. In: Proceedings of the IEEE International Conference on Fuzzy Systems, Barcelona, pp. 596–603 (2010)
14. Lucas, P.: Quality checking of medical guidelines through logical abduction. In: Coenen, F., Preece, A., Mackintosh, A. (eds.) Proceedings of AI-2003 (Research and Developments in Intelligent Systems XX), pp. 309–321. Springer, London (2003)
15. Machado, J., Abelha, A., Novais, P., Neves, J., Neves, J.: Quality of Service in healthcare units. Int. J. Comput. Aided Eng. Technol. 2, 436–449 (2010)
16. Cardoso, L., Marins, F., Magalhães, R., Marins, N., Oliveira, T., Vicente, H., Abelha, A., Machado, J., Neves, J.: Abstract Computation in Schizophrenia Detection through Artificial Neural Network based Systems. The Scientific World Journal 2015, Article ID 467178, 1–10 (2015)
17. Vicente, H., Dias, S., Fernandes, A., Abelha, A., Machado, J., Neves, J.: Prediction of the Quality of Public Water Supply using Artificial Neural Networks. Journal of Water Supply: Research and Technology – AQUA 61, 446–459 (2012)
18. Salvador, C., Martins, M.R., Vicente, H., Neves, J., Arteiro, J.M., Caldeira, A.T.: Modelling Molecular and Inorganic Data of Amanita ponderosa Mushrooms using Artificial Neural Networks. Agroforestry Systems 87, 295–302 (2013)
19. Carneiro, D., Novais, P., Andrade, F., Zeleznikow, J., Neves, J.: Using Case-Based Reasoning and Principled Negotiation to provide decision support for dispute resolution. Knowledge and Information Systems 36, 789–826 (2013)
20. Mendes, R., Kennedy, J., Neves, J.: The Fully Informed Particle Swarm: Simpler, Maybe Better. IEEE Transactions on Evolutionary Computation 8, 204–210 (2004)

Gaming and Robotics to Transforming Learning

Dalila Alves Durães

Secondary School of Caldas Taipas Guimarães, Portugal
daliladuraes@gmail.com

Abstract. Gaming and Robotics are a very motivating approach for project-based learning. This paper reports a programming (through games) and robotics-based learning in school with student aged 6 to 15. The students work in groups to programming simple task in coding; constructing and programming robots using Lego Mindstorms EV3 Kits; and create a map to move the robots avoiding obstacles.

Keywords: Coding, robotics, analytical thinking, creativity, team working.

1 Introduction

Nowadays, youth unemployment is one of Europe biggest challenges. At the same time, Europe is experiencing a growing skills gap in the ICT sector, which is expected to see a shortage of 900.000 ICT practitioners in the European labor market by the end of 2020. In addition the number of jobs requiring digital skills and competences continues to increase, while the number of computer science graduates across Europe is stagnating [6]. According to Eurostat, in 2012, approximately five and half million young people in Europe did not complete upper secondary education and did not participate in any training program [8]. This is a lost opportunity and potentially an important strain on Europe's future competitiveness. To address these challenges the European Commission would like to ensure that the children are better equipped to work and live in Digital Agenda, and improved for enhanced digital skills for all [6].

Promoting transversal competences such as critical thinking, problem solving, creativity, teamwork and communication skills development in science and technology at all levels of education, is one of the key elements within the "Innovation Union" flagship initiative under "Europe 2020". The aim of the campaign is to promote computer thinking through a mixture of online and offline, real-life activities, with a view to establishing coding as a key competence within every education system in Europe [6].

Robotics and gaming cannot just support studies in math's, science, technology, and engineering. Starting early means that the student's will be more inclined to consider computer science studies and ICT related careers.

2 Educational Robotics

The "Innovation Union" communication recognizes that weaknesses remain with science teaching. Finding of current surveys of school student's attitudes to Science

© Springer International Publishing Switzerland 2015
T. Di Mascio et al. (eds.), *Methodologies & Intelligent Systems for Technology Enhanced Learning,*
Advances in Intelligent Systems and Computing 374, DOI: 10.1007/978-3-319-19632-9_7

and Technology witness lack of interest driven by perception that Science and Technology subjects are hard, boring and "only for boys" [12].

Nowadays, there are voices in education world-wide arguing that there is a gap between the current educational practices in schools, including science and technology education, and the modern societal needs calling for an education that will foster creativity and inventiveness [4].

Research indicates that the market growth for personal robots, including those used for entertainment and educational purposes, has been tremendous and trend may continue over the coming decades [13]. However, a report from OECD remarked, "Technology is everywhere, except in schools" [10].

Educational robotics is emerged as a unique learning tool creates a learning environment that attracts and keeps student's interested and motivated with hands-on, fun learning. So educational robotics represents a powerful, engaging tool for youth learning because children can touch and directly manipulate the robots, resulting in hands-on, minds-on, self- directed learning. However, what is unique to our human-made world today is the fusion of electronics and software with mechanical structures—the discipline of robotics. This provides opportunities for young children to learn about mechanics, sensors, motors, programming, and the digital domain. With the growing popularity of robotics, the use of educational robotic kits is becoming widespread in high schools, middle schools, and elementary schools [7].

Learning with educational robotics provides students with opportunities for them to stop, question, and think deeply about technology. When designing, constructing, programming and documenting autonomous robots, students not only learn how technology works, but they also apply the skills and content knowledge learned in school in a meaningful and exciting way. Educational robotics is rich with opportunities to integrate not only STEM but also many other disciplines, including literacy, social studies, dancing, music and art, while giving students the opportunity to find new ways to work together to foster collaboration skills, express themselves using the technological tool, problem-solve, and think critically and innovatively. Most importantly, educational robotics provides a fun and exciting learning environment because of its hands-on nature and the integration of technology. The engaging learning environment motivates students to learn whatever skills and knowledge needed for them to accomplish their goals in order to complete the projects of their interest.

3   Methodology

In Portugal the childhood curriculum is focused on literacy, numeracy and science. In this paper it is described a learning experiment in an extracurricular activity named "Robotics Club". This robotics club program is explicitly designed to address "the missing middle letters" of STEM in early childhood education—the T (technology) and the E (engineering). The objective of this study is to compare the results of students from this club with other students in the same class.

This activity was designed on a constructivist approach and the curriculum introduces and uses six powerful ideas from computer science in a gaming and robotics context. It is structured to gaming, engineering design process, sequencing and control flow, loops and parameters sensors, and branches.

The students of this robotics club have age range 6 to 15. This activity has 29 students and they were separated by groups, being them: first group, students aged range 6 to 9 years old; second group students, aged range 10 and 12 years old; and third group students, aged range 13 and 15 years old. In the sessions, each group has in maximum six students.

At the heart of our robotics club is the curriculum, which consists of approximately 30 hours of training involving gaming to create coding, building and programming of robots. The format of the activities involves a short introductory presentation by an informal educator followed by hands-on activities supported by structured worksheets. Participants typically work in pairs to complete the majority of robotics tasks, and small groups of three or four students are formed for more advanced challenges.

In the ten first sessions (each session has one hour), each group programming simple tasks, using the graphics environment named Studio.Code. This environment is like a game where the students solving simple tasks like "How do I need to move to get to one destination". In this case, the students need to know, how to move and the direction to do it. When the student's finish the task, they see the results in a graphics mode, but they also see the code programing solution. In others tasks levels, the students need to solving the task and also need to simplify the algorithms using loops. The group whose age range 6 to 9 begins programming in a level one of Studio.Code. The order groups, whose age range 10 to 15, begin programming in a medium and higher level, respectively.

In the next fifteen sessions, each group constructs and programming a robot using Lego Mindstorms EV3 Kit. Sample lessons cover skills such as writing a simple program to display text on the brick, programming the robot motors for movement and various turns, using loops in a program, navigation to avoid obstacles using touch, ultrasonic and infra-red sensors; programming the gyroscope sensor to move and rotate in a certain direction; and the color sensor to track a line or to recognize same colors. No guidance was provided by the teacher, just the necessary technical support, for example how to use the programming blocks or how to make data logging and position-time graphs. Worksheets were handed to students presenting open questions/problems and offering technical support. For example: "devise a program that will make your car move backwards...", "make your car move in a linear constantly accelerated motion, write down your ideas...", "can you add some more seconds of change direction? How does the graph change now?"

In the last five sessions, the student's become familiar with RoboMind platform. This platform is a robot simulation environment program. The program is split in to parts: the left with the text editor and the right with the monitor. The text editor is where the scrip of the programming appears. The monitor is the screen where the robot and its world are visualized in. The environment can easily be explored by simply dragging the map with the mouse. The students can also zoom in order to adjust the level of detail to their preferences. The camera can track the robot automatically. The students can automatically write the programming in the editor text or drags the solution in the monitor. The Robomind platform allows the students to understanding the scrip language and develop; understanding algorithms, and turn your algorithms into code for your robot make basic sequencing problem more efficient and complicated by adding selection, repetition and procedures. Finally, the students apply these programs to Lego Mindstorms EV3 Kit and visualizing the results.

4   Results and Discussions

To identify the realization of experience helped the students in the acquisition of the concepts involved, it was applied a pre-test and post-test with the group of students where they need to solving ten different tasks.

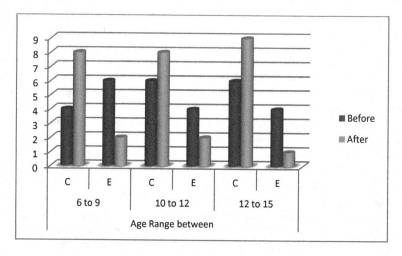

Fig. 1. The Correct (C) and the Error (E) from each group for different tasks activities

The Figure 1 graph illustrates the results obtained, consisting an improvement of 40% in the post- test hits in the age range 6 to 9; 20% in the post test hits in the age range 10 to 12; and 30% in the post test hits in the age range 12 to 15.

Comparing this students that participate in the robotics club with other students of the same classroom that not participate in the robotics classroom, we obtained the results illustrated in the Figure 2.

The results obtained by this students that participated in the robotic club comparing with the other students that didn't participated in the robotics club are better in science and technology subject. In math's, the group age 12 to 15 and the group age range 6 to 9, obtain betters results.

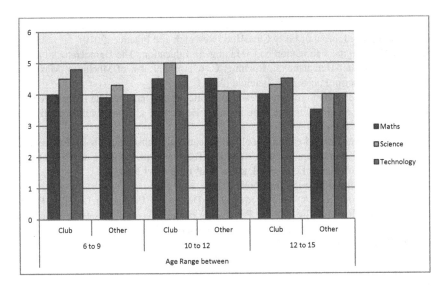

Fig. 2. Comparison Robotics Club students with other students in the same class who hasn't participated in the Robotics Club

5   Conclusions

The experiences in the classroom realized the great interest that the students had to carry out the programming in the environment so that they could use and test the robot.

This experiences show improves results in science and technology subjects. In the math's subjects the results aren't so visible. However the teachers of math's subjects refer that the students that participate in the robotics club are more motivate for the math's subjects.

In the robotics club, the students have a high range of motivation and interesting in solving new tasks. This kind of attitude has generated a climate of competitiveness, causing them to attempt to solve in less time and with less possibility of flaws in relation to colleagues.

Although, this paper offers a small-scale study and for that reason we should be cautious to draw a general conclusions from the finding. However, evidences from this robotic club activity provide positives indications that it helped the students to improve their coding abilities and they motivation.

References

1. Alimisis, D.: Computer-based modeling as learning tool in kinematics. In: Proceedings of the 6th International Conference on Technology in Mathematics Teaching, University of Thessaly, pp. 191–195 (2003)
2. Bers, M.U.: Engineers and storytellers: Using robotic manipulatives to develop technological fluency in early childhood. In: Saracho, O.N., Spodek, B. (eds.) Contemporary Perspectives on Science and Technology in Early Childhood Education, pp. 105–125. Information Age, Charlotte (2008b)

3. Bers, M.U.: The TangibleK Robotics Program: Applied Computational Thinking for Young Children. Journal of Early ChildHood Research and Practice (2010)
4. Blikstein, P.: Digital Fabrication and 'Making' in Education: The Democratization of Invention. In: Walter-Herrmann, J., Büching, C. (eds.) FabLabs: of Machines, Makers and Inventors. Transcript Publishers, Bielefeld (2013)
5. Craig, J.J.: Introduction to robotics. Pearson Prentice Hall, Upper Saddle River (2005)
6. Digital Agenda of the European Commission, Education, Culture, Multilingualism and Youth, Brussels (2014)
7. Eguchi, A.: Robotics as a Learning Tool for Educational Transformation. In: Proceeding of 4th International Workshop Teaching Robotics, Teaching with Robotics & 5th International Conference Robotics in Education Padova (Italy), July 18 (2014)
8. Eurostat, Education, Regional Yearbook (2014)
9. Kochakornjarupong, D.: A Web-based System Design for Enhancing Learning Problem Solving in Artificial Intelligence. In: Proceeding of the Seventh International Conference on eLearning for Knowledge-Based Society, Thailand, December 16-17 (2010)
10. OECD/CERI International Conference "Learning in the 21st Century: Research, Innovation and Policy", New Millennium Learners Initial findings on the effects of digital technologies on school-age learners (2008)
11. Rogers, C.B., Wendell, K., Foster, J.: The academic bookshelf: A review of the NAE Report, Engineering in K-12 education. Journal of Engineering Education 99(2), 179–181 (2010)
12. TISME, Researching Science & Mathematics Education, What influence participation in Science and Mathematics? A briefing paper from the targeted initiative on Science and Mathematics Education, TISME (2012)
13. UNEC, World Robotics 2003 – Statistics, Market Analysis, Forecasts, Case Studies and Profitability of Robot Investment is available, United Nations Office at Geneva (2003)

SmartHeart CABG Edu:
First Prototype and Preliminary Evaluation

Gabriele Di Giammarco[1], Tania Di Mascio[2], Michele Di Mauro[3],
Antonietta Tarquinio[4], and Pierpaolo Vittorini[3]

[1] Dep. of Neurosciences, Imaging and Clinical Sciences
University of Chieti, Chieti, Italy
[2] Dep. of Information Engineering, Computer Science and Mathematics
University of L'Aquila, L'Aquila, Italy
[3] Dep. of Life, Health and Environmental Sciences
University of L'Aquila, L'Aquila, Italy
[4] Cardiology Unit
AUSL Pescara, Pescara, Italy
{gabrieledigiammarco57,mdimauro1973}@gmail.com,
{tania.dimascio,pierpaolo.vittorini}@univaq.it,
a.tarquinio@tiscali.it

Abstract. The paper reports on the preliminary evaluation of the SmartHeart CABG Edu Android app. The app was conceived to be an innovative and up-to-date tool for patient education, the first of its kind in the Italian context. In particular, the app was developed to provide (i) educational material for patients about to undergo Coronary Artery Bypass Graft (CABG) surgery and (ii) a set of self-assessment tools concerning health status (i.e., BMI calculator, LDL cholesterol calculator and anxiety assessment tool) and usability (i.e., SEQ and SUS). The educational material informs patients on Coronary Artery Disease, on the CABG surgery clinical pathway and on the healthy lifestyle behaviors to implement in order to enhance self-care, while the self-assessment tools stimulate patients to check on progress in their health status. The evaluation concerning the app usability has given overall positive rating results, whereas further research with a higher number of patients is advisable in order to evaluate pre-operative anxiety more significantly.

Keywords: Pre-operative education, user experience, anxiety.

1 Introduction

Patient education is the process by which health professionals impart information to patients, their family members and/or their caregivers. Health education is one of the most important responsibilities of nurses. In comparing acute illness to chronic disease, the goals of patient education differ. During acute illness education may be considered as health information given at bedside to help patients understand diagnosis and treatment and, therefore, enhance their adherence to clinical care collaborating with the healthcare providers. In the case of chronic illness, education helps the patient and the family

© Springer International Publishing Switzerland 2015
T. Di Mascio et al. (eds.), *Methodologies & Intelligent Systems for Technology Enhanced Learning*,
Advances in Intelligent Systems and Computing 374, DOI: 10.1007/978-3-319-19632-9_8

members become knowledgeable about the disease, develop expertise in self-care and improve lifestyle by changing unhealthy behaviors and correcting risk factors.

In Coronary Artery Disease (CAD), education is essential for empowering patients in both the acute and the chronic phases of the disease, involving the patients in their plan of care in the former case and helping them being compliant to healthy behaviors, in the latter. CAD is a chronic disease and in this way, even after CABG surgery, disease progression can be slowed down in order to enable patients live a normal, healthy life. Pre-operative education facilitates the patients' understanding of the diagnosis and treatment of Coronary Artery Disease and of the consequences of their health-care decisions in adopting healthy lifestyles and improving outcomes. For those about to undergo cardiac surgery (CABG), patient education can be even more significant since these patients may experience great anxiety for a life-threatening intervention. Various studies have been conducted concerning pre-operative education of patients about to undergo CABG surgery. Geyer, Mogotlane and Young [15] as well as Black and Hawks [2] state that patients confronted with a life-threatening disease experience intense emotions such as depression, aggression, anxiety, frustration and fear which can cause them to behave irrationally. Gallagher and McKinley [5] report that patients who experience anxiety before a CABG have more post-operative pain, less long-term relief of cardiac signs and symptoms, more re-admissions, poorer quality of life and worse long-term psychological outcomes. Therefore, it is one of the responsibilities of nurses to help the patients to manage their stress and anxiety, as they are in close contact with those patients [14]. One research study concluded that patient education before CABG surgery will not only reduce patient's anxiety but also the need for sedatives; the authors though suggest further studies with larger samples [10]. A study on nursing care of the CABG patient highlighted the importance of appropriate timing of pre-operative education [8]. On the contrary, other research findings demonstrate that there is no benefit to be gained from pre-operative education [13].

In today's information technology dominated world, the use of smartphones, tablets and other mobile devices are getting us used to the fact that there is an "app" for just about anything. At the same time, patients have become more active partners in their clinical care and decision making. In such a context the use of mobile devices has become of increasing importance for searching and disseminating health information and has turned mobile learning into an everyday activity. Such devices are easily customized and can support individual learning in all places: at home, at work, at school, in hospital, etc. For this reason the SmartHeart CABG Edu app was conceived; it is meant to be an innovative and up-to-date tool for patient education, the first of its kind in the Italian context.

2   SmartHeart CABG Edu

The SmartHeart CABG Edu is an app that was developed to provide educational material for patients about to undergo CABG surgery and a set of self-assessment tools concerning both their health status (i.e., BMI calculator [6], LDL cholesterol calculator [4] and anxiety assessment tool [16]) and their experience with the app, i.e., Single Ease Question (SEQ) [11] and System Usability Scale (SUS) [3]. The app runs on Android[1],

---

[1] Minimum version required is 4.0.3 API 15.

was developed using Android Studio, is 2.2Mb in size and does not require Internet to access the educational resources.

The app is divided in two main sections: the educational material and the self-assessment tools. Tapping the SmartHeart icon will have the user see the main page, from which access to each of the two sections is possible (Fig. 1-a). Tapping on the educational material button, the user is taken to the educational content which is organized in four subsections: anatomy of the heart and CAD info, diagnostic tests and therapeutic procedures, CABG surgery, cardiac rehab and healthy lifestyles (Fig. 1-b). Tapping on the self-assessment tools button, the user is taken to the respective page. Here we find: the BMI and the LDL cholesterol calculators, the anxiety assessment tool and the two usability evaluation tools (Fig. 1-c).

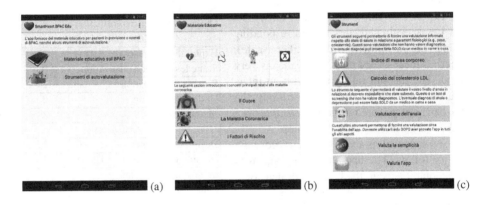

Fig. 1. (a) App's main page showing both sections available, (b) educational material divided into four subsections indicated by clipart icons, (c) self-assessment tools page

3   Evaluation

The choice of performing one type of evaluation over another was established in relation to the stage of the project, the material to evaluate, the experts involved in the evaluation, as well as the time constraints and the available resources of the project [9]. Since our project is in its first release, we decided (i) to ask one available usability expert for an heuristic evaluation to generate large numbers of potential usability problems (Subsect. 3.1); (ii) to inquire both primary and secondary stakeholders, i.e., patients and health-care workers, respectively (Subsect. 3.2). Finally, we investigated on the pre-operative anxiety of patients, before or after their reading the educational material (Subsect. 3.3).

3.1   Expert-Based Review

The most popular type of expert evaluation is the heuristic evaluation [9]. The evaluator, in our case one of the paper's authors, used a set of heuristics to guide the evaluations and rate the potential problems in terms of their importance.

Starting from its original proposal by Nielsen [9], heuristic evaluation has been adapted in many ways. In the case of smartphone apps, for example, a recent study by Gomez et al. [7] reported a new check list to evaluate mobile interfaces. This new check list reuses 69% of literature heuristics; the rest deriving from best-practices and recommendations for mobile interfaces.

Given that the app is a first prototype, the heuristics used for the evaluation were reduced and limited to only those actually applicable. Accordingly, Table 1 summarizes the expert-evaluation results: the table lists the items evaluated and the corresponding rating, i.e., a numeric value (Likert scale), where 0 means very poor and 4 means very good. The median rating is 2 with an interquartile range (IQR) of 2^2. Accordingly, the app can be considered *sufficiently usable* even if the ratings have a large dispersion.

Few suggestions about how to improve usability follow. The app should be improved in adaptability (e.g. the buttons' dimensions) and in the user personalization (e.g., an elderly person might have problems with font size). The contents should be supported via multimedia, in order to justify a technological support. Visual design might be improved (e.g., using icons with text labels or using a coherent color palette). The affordance of browsing and navigation might be improved by introducing a menu (e.g., search the educational material). Finally, the auto-evaluation questionnaires' layout should be presented using boxes in order to improve the text comprehension.

Table 1. Expert-based evaluation result

Item	Rate
Visibility of system status	
System status feedback	3
Response time	4
Selection/input of data	2
Presentation adaptation	0
Match between system and the real world	
Metaphors/mental models	2
Navigational structure	1
Menus	1
Simplicity	2
Output of numeric information	4
User control	
Explorable interfaces	2
Some level of personalization	0
Process confirmation	2
Undo/cancellation	2
Menus control	2
Consistency	
Design consistency	3
Naming convention consistency	2
Menus/task consistency	3
Functional goals consistency	3
System response consistency	2
Orientation	4
Flexibility and efficiency of use	
Search	0
Navigation	2
Aesthetic and minimalist design	
Multimedia content	1
Icons	2
Menus	1
Orientation	3
Navigation	1

3.2 User Experience

For investigating the user experience, we involved both primary and secondary stakeholders. The primary stakeholders are patients that have already undergone, or are about

---

[2] Median and IQR are the most proper indexes for central tendency and dispersion for categorical variables. Mean and standard deviation are instead more suitable in case of numeric variables.

to undergo CABG surgery. The secondary stakeholders are the health-care workers of the involved hospital wards (e.g., critical care unit, cardiac surgery).

In order to learn about the users' reactions to our app in general and about the preliminary usability issues carried out from the expert based evaluation in particular, we decided to perform personal observational evaluation sessions (i.e., involving real users that are observed when performing tasks with the system) and a final focus group (i.e., a moderated discussion that typically involves 5 to 10 participants). Quantitative and qualitative data were gathered during the entire evaluation. Qualitative data was annotated by the evaluator. Quantitative data regarded their user experience, and was reported through the SEQ [11] and the SUS [3] scores, and the Time-On-Task [12]. As known, the Single Ease Question (SEQ) is a 7-point rating scale to assess how difficult users find a task. It is administered after a user attempts a task, by asking the simple question: "How difficult or easy was the task to complete?". The higher the value of the response, the easier the task is. As for the SEQ, we asked the stakeholders to rate the three tasks of: (T1) reading the educational material, (T2) using the BMI calculator, and (T3) using the LDL cholesterol calculator. The System Usability Scale (SUS) instead is a reliable tool for measuring usability, consisting of a 10 item questionnaire with five response options per item. It can be used to evaluate a wide variety of products and services, including mobile apps. The tool rates a system with a score ranging from 0 to 100. The higher the score, the more usable the system is. For a simpler interpretation of the SUS score, we introduced a qualitative interpretation, based on the thresholds computed in [1], that "converts" the score into the seven adjectives for the app, namely: "worst imaginable", "awful", "poor", "OK", "good", "excellent", "best imaginable". Finally, the Time-On-Task (ToT) is the total time taken by a user to complete a task.

3.2.1 Primary Stakeholders

The evaluation was carried out from the 23rd of January, 2015 to the 2nd of February, 2015. Before the observational sessions, the evaluator explained the goals of the evaluation to the patients, using the guidelines for an elderly observational evaluation where necessary (i.e., speaking aloud and slowly, using easy vocabulary and describing the goal using a family approach, to make sure all patients, even of different ages, understood perfectly). A friendly environment was established. The focus groups took place after the observational evaluation sessions. A total of 10 patients were involved in the study, two coming from the Coronary Unit of Pescara Hospital, five from the Neurosurgery Unit of Chieti Hospital, three from the Cardiology Unit of Avezzano Hospital. Not all patients completed the SEQ and SUS questionnaires, but for all of them we could analyze the ToT data. The results follow.

SEQ The distribution of the SEQ scores is reported in Fig. 2. In terms of central tendency, the median for T1 is "easy" (IQR is ["average", "very easy"]), for T2 is "easy" (IQR is ["not difficult", "easy"]), for T3 is "easy" (IQR is ["average", "easy"]). Therefore, on average, no particular difficulties are reported, even if few users rated all tasks below the sufficiency.

SUS The distribution of the SUS score is shown in Fig. 3. The median rating is "OK". To investigate further into the reasons of the score, we also analyzed the answers to each question of the SUS score (Fig. 4). There are no questions with clearly negative answers. However, there are two questions with discordant answers, i.e., questions "D1" (I think that I would like to use this system frequently) and "D10" (I needed to learn a lot of things to use this system). As for "D1", the reason is that the app was developed for the pre-operative phase, and not for daily use. The results regarding "D10" can be easily explained by the fact that, since the patients were elderly people, they were not accustomed to the use of smartphones. On the other hand, the more positive aspects were in connection with "D3" (I thought the system was easy to use), "D5" (I found the various functions in this system were well integrated) and "D7" (Most people would learn to use this system very quickly).

ToT The results of the descriptive analysis are listed in Table 2. The four tasks of completing the anxiety questionnaire ("Anxiety" task), the BMI calculation ("BMI" task), the LDL estimation ("Cholesterol" task) and reading a page of the educational material ("Fact" task) are in the first column, the corresponding average seconds taken are listed in the second. The average times do not highlight particular difficulties in completing the tasks.

Table 2. ToT descriptive analysis

Task	Mean
1 Anxiety	68.24
2 BMI	22.35
3 Cholesterol	61.53
4 Fact	52.84

By observing the users during their interactions, the system appeared in general "browsable": no user asked for support in browsing the GUI, however the time spent to individuate the buttons was in general high wrt standard users; this is not surprising considering that the involved users were elderly people. Users also asked about the possibility of having the educational material in paper format: it is natural given that the users were elderly people. Finally, no particular problems emerged while calculating the body mass index and the LDL cholesterol. In all cases, improvements in terms of a bigger font size and button affordance (e.g., by adding text together with icons) was considered as required.

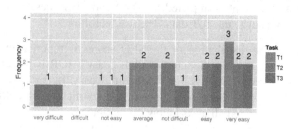

	T1	T2	T3
very difficult	0	0	1
difficult	0	0	0
not easy	1	1	1
average	2	0	0
not difficult	0	2	1
easy	1	2	2
very easy	3	2	2

Fig. 2. Descriptive analysis of the SEQ score

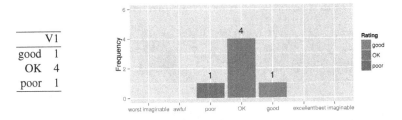

	V1
good	1
OK	4
poor	1

Fig. 3. Descriptive analysis of the SUS scores

The focus groups confirmed some issues carried out in the observational evaluation phase, such as the font size issue, but also highlighted the enthusiasm of users in reading the educational material. One user thanked the evaluator for having provided him with educational material on CABG surgery, another asked for a paper copy to bring to her own cousin (so to better explain her problems). Furthermore, users suggested to improve on the multimediality of the material (e.g., by adding videos). A final remark is on the personalization of the app: users appreciated the possibility to calculate their own data and asked for more functions supporting their personal health status, e.g., storing their blood pressure values, alarms that reminded medication-intake time.

3.2.2 Secondary Stakeholders
A total of 38 secondary stakeholders were involved in the study. The results follow, divided into the SEQ and the SUS scores.

SEQ The results of the descriptive analysis are in Fig. 5. For each task, the histogram shows the ratings in terms of distribution, while the table reports on the actual values. As can be noticed, the majority of the secondary stakeholders rated each task as "very easy" to accomplish.

SUS The results of the descriptive analysis are in Fig. 6. In summary, the app is considered in the majority of the cases as the "best imaginable".

Brainstorming with the secondary stakeholders highlighted two possible improvements in the usability of both the BMI calculator and the educational material browser. As a result, the BMI calculator was made to accept both centimeters or meters for inserting the patients' height. As for the educational material, to improve its readability, access to the subsections was facilitated by making them all visible simultaneously, instead of having to swipe across the screen in order to see them.

3.3 Pre-operative Anxiety

The analysis of the anxiety level was limited only to primary stakeholders that were about to undergo CABG surgery. To such patients, we submitted the Hospital Anxiety and Depression Scale (HADS) [16], which was actually developed to "[...] facilitate the large task of detection and management of emotional disorder in patients under investigation and treatment in medical and surgical departments [...]". Eight patients were

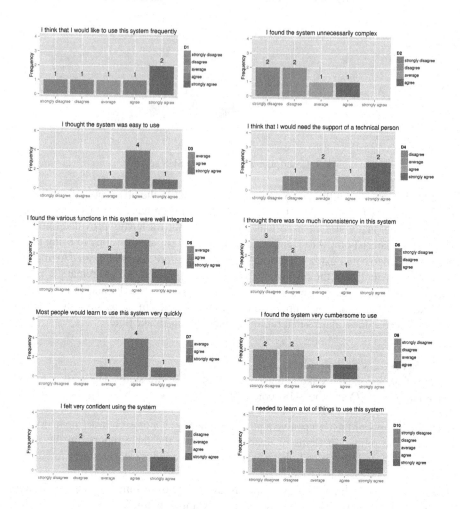

Fig. 4. Descriptive analysis of the SUS score by question

enrolled: they were randomly asked to fill the questionnaire either before or after having read the educational material. Fig. 7 summarizes the descriptive analysis of the data: on the left the frequency table, on the right the corresponding histogram. No patients reported significant anxiety levels after having read the educational material, vs two that instead showed a probable anxious status and two that did not reveal anxiety. Such a difference was not statistically significant (Wilcoxon rank-sum test, $p = 0.1814$), but given the reduced number of patients enrolled in the study and that the p-value is close to the significance, we can still consider such a result as encouraging.

	T1	T2	T3
very difficult	0	0	1
difficult	0	0	0
not easy	0	0	0
average	2	1	1
not difficult	3	4	5
easy	6	10	3
very easy	27	23	28

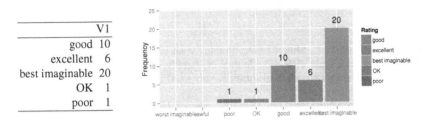

Fig. 5. Descriptive analysis of the SEQ score

	V1
good	10
excellent	6
best imaginable	20
OK	1
poor	1

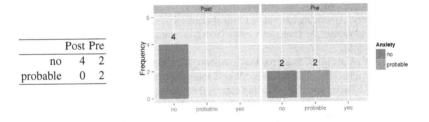

Fig. 6. Descriptive analysis of the SUS score

	Post	Pre
no	4	2
probable	0	2

Fig. 7. Descriptive analysis of the anxiety levels

4 Conclusions

The paper reported on the first prototype of an Android app developed for providing educational material to patients about to undergo CABG surgery. It briefly describes the app, then focuses on its evaluation from the twofold perspective of usability and health outcomes.

As for usability, the evaluations coming from the expert concluded that the app user interface appears sufficiently usable and highlighted several points that can be improved. The primary stakeholders did not reported major usability problems with the app, but highlighted – both via quantitative and qualitative data – some improvements that can be introduced. On the other hand, the secondary stakeholders evaluated the app with very good ratings: the main tasks were all considered very easy to accomplish, and the app was considered – in the majority of the cases – as the best imaginable.

In terms of anxiety, given the reduced number of patients, we could not find a statistically significant difference between patients that read and didn't read the educational material. However, given the reduced number of enrolled patients and the p-value close to the significance, we can still consider the result as encouraging.

Our future work will focus on improving the app usability according to the evaluation results and extending the experiment regarding anxiety by enrolling a larger sample of patients.

Acknowledgment. We would thank the personnel working in the Coronary Unit of the Pescara Hospital, in the Neurosurgery of Chieti Hospital and in the Cardiology Unit of the Avezzano Hospital for their cooperation.

References

1. Bangor, A., Kortum, P., Miller, J.: Determining what individual SUS scores mean: Adding an adjective rating scale. JUS - The Journal of Usability Studies 4(3), 114–123 (2009)
2. Black, J.M., Hawks, J.H.: Medical-Surgical Nursing: Clinical Management for Positive Outcomes (2009)
3. Brooke, J.: SUS-a quick and dirty usability scale. Usability Evaluation in Industry 189, 194 (1996)
4. Friedewald, W.T., Levy, R.I., Fredrickson, D.S.: Estimation of the concentration of low-density lipoprotein cholesterol in plasma, without use of the preparative ultracentrifuge. Clinical Chemistry 18(6), 499–502 (1972)
5. Gallagher, R., McKinley, S.: Stressors and anxiety in patients undergoing coronary artery bypass surgery. American Journal of Critical Care 16(3), 248–257 (2007)
6. Garrow, J.S., Webster, J.: Quetelet's index (w/h2) as a measure of fatness. International Journal of Obesity 9(2), 147–153 (1984)
7. Gómez, R.Y., Cascado Caballero, D., Sevillano, J.L.: Heuristic evaluation on mobile interfaces: A new checklist. The Scientific World Journal 2014, e434326 (September 2014)
8. Martin, C.G., Turkelson, S.L.: Nursing care of the patient undergoing coronary artery bypass grafting. Journal of Cardiovascular Nursing 21(2), 109–117 (2006)
9. Nielsen, J., Mack, R.L.: Usability Inspection Methods. John Wiley & Sons Inc., New York (1994)
10. Rahimianfar, A.A., Javadi, S.S., Dehghani, H., Sareban, M.T., Akbarzade, T., Eslami, A., Rahimianfar, F., Khosravi, A.: The effect of training booklet on anxiety level of the patients candidate for coronary artery bypass graft surgery. Journal of Biology and Today's World 2(10) (2013)
11. Sauro, J., Dumas, J.S.: Comparison of three one-question, post-task usability questionnaires. In: Proceedings of the SIGCHI Conference on Human Factors in Computing Systems, pp. 1599–1608. ACM (2009)
12. Seffah, A., Donyaee, M., Kline, R.B., Padda, H.K.: Usability measurement and metrics: A consolidated model. Software Quality Journal 14(2), 159–178 (2006)
13. Shuldham, C.M., Fleming, S., Goodman, H.: The impact of pre-operative education on recovery following coronary artery bypass surgery. a randomized controlled clinical trial. European Heart Journal 23(8), 666–674 (2002)
14. Towell, A., Nel, E.: Pre-operative education programme for patients undergoing coronary artery bypass surgery. Africa Journal of Nursing and Midwifery 12(1), 3–14 (2010)
15. Young, A., Geyer, N., Young, A.: Juta's manual of nursing, No. v. 1. Juta, Limited (2009)
16. Zigmond, A.S., Snaith, R.P.: The hospital anxiety and depression scale. Acta Psychiatrica Scandinavica 67(6), 361–370 (1983)

Influence of Gaming Activities
on Cognitive Performances

Maria Rosita Cecilia, Dina Di Giacomo, and Pierpaolo Vittorini

Dep. of Life, Health and Environmental Sciences
University of L'Aquila
L'Aquila, Italy
mariarosita.cecilia@graduate.univaq.it,
{dina.digiacomo,pierpaolo.vittorini}@univaq.it

Abstract. Playing games is an important voluntary activity that promotes cognitive, social and emotional development. In addition to traditional games, the advent of new technologies has favored an explosion of computer games, very popular among children. Against this background, the authors report their investigation regarding the effect of gaming activities on the cognitive performances of 7-11 years old children. The aim of the research was to analyze if both computer and traditional games had a positive influence on cognitive performance in childhood. 67 students participated in the study. The BVN 5-11 neuro-psychological test battery and an experimental questionnaire to detect the using habits of boards and technological games were used. Our findings highlighted that both the traditional and technological stimulation resulted effective in improving the cognitive performances of children.

Keywords: Cognitive performances, board games, technological instruments.

1 Introduction

Playing games is an important voluntary activity that promotes cognitive, social and emotional development. It is intrinsically motivating and it represents a powerful mediator for learning throughout a person's life [10]. At school, playing games helps students adjust to its environment, thereby fostering engagement, and enhances children' learning readiness, learning behaviors, and problem-solving skills [7]. However, the characteristics of playful activities changed in the last decade, becoming increasingly complex. More precisely, the advent of new technologies has favored an explosion of computer games, which are extremely appealing to children and adolescents. Considering the educational function of playing games [9], the fast diffusion of computer technology has also proposed many changes in the field of learning [1]. Indeed, nowadays, computer is a supportive cognitive tool for learning, to achieve specific pedagogical goals [12]. On the other hand, the design of technological tools has become a source of study for educational researchers and instructional designers investigating how various aspects of game design may support cognitive performances. For example, it is observed that adventure games encourage inferential and proactive thinking [8]. Some studies also suggest that computer game training enhances cognitive performance on tasks other

© Springer International Publishing Switzerland 2015
T. Di Mascio et al. (eds.), *Methodologies & Intelligent Systems for Technology Enhanced Learning*,
Advances in Intelligent Systems and Computing 374, DOI: 10.1007/978-3-319-19632-9_9

than those specific to the game. For example, expert technological game players often outperform non-players on measures of basic attention, memory and executive control [3]. Computer gaming may also provide an efficient training regimen to induce a general speeding of perceptual reaction times without decreases in accuracy of performance [6]. Moreover, games can produce engagement and delight in learning, offering a powerful format for educational environments [4]. As a result, many researchers have developed games for educational purposes, for example [5]. However, not all educators and parents are convinced that educational computer games can be beneficial to students and, for this reason, the attempts to create educational games have not reached schools yet [11].

Against this background, the authors report in the paper their investigation regarding the effect of gaming activities on the cognitive performances of 7-11 years old children. The aim of the research was to analyze if both computer and traditional games had a positive influence on cognitive performance in childhood. In particular, the authors wanted to assess the existence of different effects between traditional and technological stimulation on cognitive performance.

2   Materials and Methods

2.1   Participants

The study was carried out in 3 months, in an public primary school in middle Italy. The sample was composed of n. 67 children in the age range of 7-11 years old (mean age = 8.7, s.d. = 1.1). No children showed neurological and/or cognitive deficits. Written informed consensus was requested to parents.

Inclusion criteria was all students 7-11 aged. Exclusion criteria were as follows:

- inadequate knowledge of Italian language;
- lack of informed consent;
- diseases or health conditions that did not allow the assessment of cognitive performances.

2.2   Tests

The authors submitted both the BVN 5-11 neuro-psychological test battery [2] and an experimental questionnaire to detect the using habits of (a) boards games and (b) technological instruments (PC, tablet, electronic games). The neuro-psychological test was submitted to the children sample, while the experimental questionnaire was proposed to their parents. Children were evaluated in school time, while parents completed the questionnaire at home.

BVN. The BVN 5-11 (acronym of "Batteria di valutazione neuropsicologia per l'età evolutiva") is a neuropsychological test battery for 5 to 11 year old children. The battery is usually used to investigate the cognitive development: we applied the sub-tests of attention, programming skills, naming and grammatical abilities, described below:

Naming test	The test was composed of figures and the child was asked to pronounce the name of 20 pictures (for example, an elephant, an onion, a shoe), giving one point for each correct answer (scores range: 0-20).
Grammatical test	The test measured the ability to identify the figure corresponding to the sentence pronounced by the examiner (for example, "The boy is running") . The test has 18 items (scores range: 0-18).
Programming skills	The test consented to evaluate the logical reasoning ability to elaborate visuo-perceptive information. The examiner shows the child pictures with painted three different colored balls (red, blue and green balls) arranged on three pegs. The examiner also gives to the child the tool (the Tower), with the balls arranged in an initial standard position. The child looks at both the picture and the tool simultaneously. Then the child has to make the arrangement of balls identical to that of the picture, with a maximum number of moves and respecting certain rules (for example, the child can move a ball at a time). There are 12 items (scores range: 0-12).
Attention test	The test measured the sustained attention ability. The test provides a sheet with 10 rows. On each row there are small squares. In the upper part of the sheet there is the "square-target", which is repeated in the rows of the sheet. The child should search the target in each row, within one minute. The examiner gives a point for each target identified (scores range: 0-12).

Experimental Questionnaire. An ad-hoc questionnaire (self-report) was carried out to detect the using of board games' and technological instruments using of the children in their home. The experimental questionnaire was made up of two main sections. The first regards the socio-demographic characteristics of the child and of his/her parents (e.g., child age, mother qualification, father profession). The second contains seven questions regarding the traditional (board games) and innovative (technological instruments) stimulation in terms of typology preferred with related intensively using. The administration time was 20 minutes.

3   Results

The raw scores were elaborated through the STATISTICA program. The α value was set to 0.05. MANOVA 2x4 was conducted on sample's cognitive performance (4 cognitive test) and board games and technological instruments exposition.

The sample was divided:

- in terms of board games' use, in three groups: (1) Light Use, LU (= most 1 hour in a week), (2) Moderate Use, MU (= 3 hours a week), and (3) Intensive Use, IU (= more than 4 hours a week).
- in terms of technological instruments' use, in three groups: (1) No Autonomy, NA (=the child use the technological instruments only with a parental support), (2)

Basic Autonomy, BA (=the child is able to use some applications/tools only with a parental supervision), and (3) Light Autonomy, LA (=the child is able to use the applications/tools without any parental supervision).

The MANOVA showed a significant difference both in board games ($F(8; 122) = 4 : 04, p < 0.001, \eta^2 = 0.20$) and in technological instruments ($F(8; 122) = 4.62, p < 0.001, \eta^2 = 0.17$); no significant interaction between them.

The post-hoc analysis (Tukey test) evidenced the following significant differences:

– in naming, between the IU group wrt all others ($p < 0.04$);
– in grammatical, between any group wrt any other ($p < 0.01$);
– in programming skills, between any group wrt any other ($p < 0.04$).

No significant differences were found in attention. Fig. 1 shows the children cognitive performances by intensity in using board games.

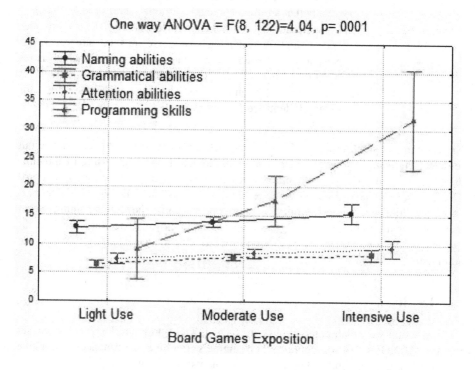

Fig. 1. Children performance by intensity in using board games

The post-hoc analysis (Tukey test) verified that the LA group is significantly different than BA and NA groups both in naming ($p < 0.001$), grammatical ($p < 0.03$), attention ($p < 0.01$) and programming skills ($p < 0.001$). Fig. 2 summarizes the cognitive performance by use of technological instrument.

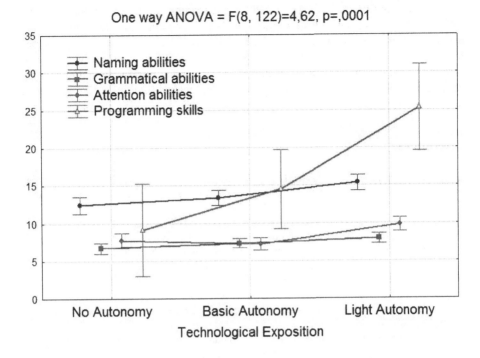

One way ANOVA = F(8, 122)=4,62, p=,0001

Fig. 2. Children performance by autonomy in using technological instruments

4   Discussion and Conclusions

Aim of the study was to investigate the influence of two different educational stimulations on cognitive performance. Our results showed that the cognitive performance of children in developmental age are improved by both board games and technological instruments, but with differences. The board games represent classical but still effectual cognitive stimulation and their effect were detected. The technological instruments' use represents a new and more articulated way to stimulate and improve the cognitive performance. Indeed, the results demonstrated that the technological instruments' and board games use have each positive effects on cognitive performance of children, improving the naming, grammatical, programming skills; the attention ability was appeared to have a major benefit by technological stimulation. Furthermore, the autonomy in the use of the technological tools and/or applications represents a good practice to improve the learning abilities in developmental age. In particular, the technological exposition in childhood can favor a better cognitive flexibility and enhanced learning. Moreover, our results highlighted the added value of the technological stimulation not only in visuo-perceptive abilities (visual tasks: programming skills and attention ability) but also in reading abilities (verbal tasks: grammatical and naming). Our results suggest that the reading abilities may have been improved by technological games because they are characterized by complex cognitive activation based on programming and logical reasoning; the effect is the cognitive flexibility influences the learning processing.

These results confirm our recent research [5]: the reading ability, and in particular the silently reading, can benefit by appropriated tools and can improve the learning processing. Furthermore, the use of technological instruments and associated tools/applications in childhood may represent a protective factor: the technological educative tools can support the children in their developmental process (a) by improving and enhancing learning abilities and (b) by favoring the scholar successes. In brief, such a protective factor may encourage the cognitive strengths and reduce the weaknesses of the child.

References

1. Amory, A., Naicker, K., Vincent, J., Adams, C.: The use of computer games as an educational tool: identification of appropriate game types and game elements. British Journal of Educational Technology 30(4), 311–321 (1999),
http://dx.doi.org/10.1111/1467-8535.00121
2. Bisiacchi, P.S., Cendron, M., Gugliotta, M., Tressoldi, P.E., Vio, C.: BVN 5-11. Batteria di valutazione neuropsicologica per l'et à evolutiva. Centro Studi Erickson, Gardolo, Trento (March 2005)
3. Boot, W.R., Kramer, A.F., Simons, D.J., Fabiani, M., Gratton, G.: The effects of video game playing on attention, memory, and executive control. Acta Psychologica 129(3), 387–398 (2008), http://www.sciencedirect.com/science/article/pii/S0001691808001200
4. Boyle, T.: Design For Multimedia Learning, 1st edn. Prentice-Hall, London (1997)
5. Di Giacomo, D., Cofini, V., Di Mascio, T., Cecilia, M.R., Fiorenzi, D., Gennari, R., Vittorini, P.: The silent reading supported by adaptive learning technology: influence in the children outcomes. Computers in Human Behavior (2014)
6. Dye, M.W., Green, C.S., Bavelier, D.: Increasing Speed of Processing With Action Video Games. Current Directions in Psychological Science 18(6), 321–326 (2009),
http://www.ncbi.nlm.nih.gov/pmc/articles/PMC2871325/
7. Ginsburg, K.R., American Academy of Pediatrics Committee on Communications, American Academy of Pediatrics Committee on Psychosocial Aspects of Child and Family Health: The importance of play in promoting healthy child development and maintaining strong parent-child bonds. Pediatrics 119(1), 182–191 (2007)
8. Pillay, H.: An investigation of cognitive processes engaged in by recreational computer game players: Implications for skills of the future. Journal of Research on Technology in Education 34(3), 336–350 (2002)
9. Pivec, M.: Editorial: Play and learn: potentials of game-based learning. British Journal of Educational Technology 38(3), 387–393 (2007),
http://dx.doi.org/10.1111/j.1467-8535.2007.00722.x
10. Rieber, L.P.: Seriously considering play: Designing interactive learning environments based on the blending of microworlds, simulations, and games. Educational Technology Research and Development 44(2), 43–58 (1996),
http://dx.doi.org/10.1007/BF02300540
11. Virvou, M., Katsionis, G., Manos, K.: Combining Software Games with Education: Evaluation of its Educational Effectiveness. Educational Technology & Society 8(2), 54–65 (2005)
12. Vosniadou, S.: International perspectives on the design of technology-supported learning environments. Routledge (1996)

Boosting Learning: Non-intrusive Monitoring of Student's Efficiency

Sérgio Gonçalves[1], Manuel Rodrigues[2], Davide Carneiro[3],
Florentino Fdez-Riverola[1], and Paulo Novais[3]

[1] Informatics Department, University of Vigo
Ourense, Spain
sgoncalves@alumnos.uvigo.es, riverola@uvigo.es
[2] Escola Superior de Tecnologia e Gestão
Felgueiras, Portugal
mfsr@estgf.ipp.pt
[3] Algoritmi Centre/Department of Informatics, University of Minho
Braga, Portugal
{dcarneiro,pjon}@di.uminho.pt

Abstract. Keeping students interested and motivated is perhaps one of the most difficult and traditional tasks assigned to teachers. With technology being engaged increasingly into learning activities, with its advantages and disadvantages, some new aspects need to be considered. Undoubtedly, technology acts as an enhancer for learning, opening new paths for teaching. However there are some drawbacks too. Keeping students in the right track, doing what they are expected to do, with commitment and motivation, becomes an enormous challenge when an amazing digital world full of all kind of temptations is at the distance of their personal smartphones or even in the computer they use to study. This excess of stimuli and the process of switching and choosing between them has as potential effects on attention, stress and mental fatigue. Stressed or fatigued students fail to deliver the required performance for the task they are engaged in. This paper presents a non-intrusive approach for monitoring student's performance in real time and measure the effect of these external variables on students. The long-term goal is to empower teachers with valuable information about the students' state, allowing them to better manage their students and teaching methodologies.

Keywords: e-learning, Fatigue, Stress, Recommendation System, Monitoring.

1 Introduction

Modern society needs to be constantly fed with new knowledge, putting an enormous amount of pressure into the formation/requalification of their citizens. Resources to education/training are finite and must be used in the most efficient way. Technology arises as a way to enhance this learning/teaching processes, providing new ways to achieve better results, and overcoming some known constraints such as qualified instructors availability, time restrictions, and individual

© Springer International Publishing Switzerland 2015
T. Di Mascio et al. (eds.), *Methodologies & Intelligent Systems for Technology Enhanced Learning,*
Advances in Intelligent Systems and Computing 374, DOI: 10.1007/978-3-319-19632-9_10

monitoring. When using technology-enhanced learning, some drawbacks need to be carefully analysed. When a student engages into an electronic course, the interaction between student and teacher, without all its non-verbal interactions, is poorer. Thus the assessment of feelings and attitudes by the teacher becomes more difficult. In that sense, the use of technological tools for teaching, may represent a risk as a significant amount of context information is lost. Since students' effectiveness and success in learning is highly related to their mood while doing it, such issues should be taken into account when in an e-learning environment. Stress [6,15], fatigue or attention, in particular, can play an important role in education [1,2]. In that sense, its analysis in an e-Learning environment, in which no contextual cues exist for the teacher to use, assumes greater importance.

Attention can be seen as a basilar cognitive process in what concerns learning. It is strongly connected with learning and assimilating new concepts, and with the quality with which this is done. The lack of attention can thus be very problematic [7] in the learning context. Moreover, prolonged attention in a challenging cognitive task for a long period of time may result stressful (especially if there are feelings of frustration) or fatiguing (especially if the task is challenging, boring or simply not appealing). This, in turn, affects motivation to continue working on the task at hand [4]. A direct relationship between attention (or distractions) and stress and fatigue can thus be established, which impacts the quality of the learning process.

While in a traditional classroom, the teacher can act on specific students to minimize signs of distraction. He can do so because he sees students and their behaviours. When in an e-learning environment this results significantly more difficult. Hence the need for appropriate remote and real-time monitoring solutions. This is traditionally done through the use of physiological sensors. However, most of the times this results uncomfortable or even impracticable, and has significant costs.

In this paper we address the use of behavioural biometrics, specifically keystroke and mouse dynamics [12], to feed machine learning techniques that can be used to distinguish scenarios in which the user shows signs of fatigue, distraction or stress. This approach can be deemed both non-invasive and non-intrusive as it relies solely on the observation of the individual's use of the mouse and keyboard [11]. It allows the teacher to observe important changes in the interaction patterns of students, which can be sign of the effect of external factors. It provides data in real time that may allow the teacher to act on students in a personalized manner, improving the efficiency of teaching methodologies.

2   A Dynamic Approach to Monitor Students' Efficiency

In the introductory section, well-known problems that may arise from fatigue, stress or distraction were briefly addressed. To cope with these problems, we propose a constant monitoring of the students in terms of their interaction patterns with the computer, combined with questionnaires and similar tools in a preliminary phase so that machine learning algorithms can be properly trained.

These interaction patterns occur mainly through the use of the keyboard and mouse. Therefore, these are the peripherals on which monitoring takes place. The use of these peripherals, as addressed in [8,9] allows to acquire contextual features that describe the interaction patterns of the user with the computer. These features reflect the behaviour and the performance of the user and how it changes under certain conditions, such as when the user is fatigued, stressed or distracted.

To implement this approach and acquire valuable information from these peripherals, a log tool was developed at the ISLab, in the University of Minho. Monitoring is implemented through the log of certain system events. Table 1 summarizes these events.

Table 1. Mouse and keyboard events registered by the log tool

Event	Description
MOV	mouse movement, in a given time, to coordinates (posX, posY)
MOUSE_DOWN	mouse button pressed down, in a given time. (left or right) and mouse position (posX, posY)
MOUSE_UP	an event similar to previous one but describing the second part of the click, when the mouse button is released
MOUSE_WHEEL	this event describes a mouse wheel scroll of a certain amount, in a given time
KEY_DOWN	identifies a given key from the keyboard being pressed down, at a given time
KEY_UP	describes the release of a given key from the keyboard, in a given time

From these events describing the interaction of the user with the mouse and the keyboard, it is possible to extract interaction features such as mouse velocity or acceleration, writing speed, time between clicks, time between keys, click duration, among others. Previous work on this data collection tool and analysis can be found in [5,10] where a more detailed analysis about this process is provided, as well as a description of all the interaction features.

The most important aspect about this approach is that it allows obtaining a measure of the user's performance (e.g. an increased distance between clicks or sum of angles represents a decreased performance). Once information about the individual's performance exists in these terms, it is possible to start monitoring the effects of fatigue (such as in prolonged periods of study or work), stress (such as in high-stake exams) or attention (such as when students are using applications that are not necessary for the assigned task), in real-time. This makes this approach especially suited to be used in e-learning environments in which students use computers, since it requires no changes in their working routines. This is the main advantage of this work, especially when compared to more traditional approaches that still rely on physiological sensors or on the availability of human experts.

2.1   The Monitoring Tool

The tool developed to acquire data from working or studying routines compiles information from the students' interaction with the mouse and keyboard, which act as sensors. The proposed framework includes not only the sheer acquisition and classification of the data, but also a presentation tier that will support the human-based or autonomous decision-making mechanisms that are now being implemented. It is a layered architecture, with information flowing as depicted in Figure 1.

The *Data Acquisition* layer is responsible for capturing information describing the behavioural patterns of the user, receiving data from events fired from the use of the mouse and the keyword. This layer encodes each event with the corresponding necessary information (e.g. timestamp, coordinates, type of click, key pressed).

The *Data Processing* layer takes the log of events generated by the Data Acquisition layer and transforms them into the set of interaction features previously mentioned. In this stage outlier values are also filtered to improve the quality of the data.

The *Classification* layer is where the interaction features are used, together with the results of questionnaires and other tools, to train models of interaction. These allow to later classify the behaviour of students in real-time.

The *Data Access* layer, is responsible for providing access to the different levels of data mentioned before (e.g. raw data, features, models). It also provides access to historic data about each user, which is very important to establish a profile and perform long-term analyses (e.g. study the evolution of performance throughout a semester).

Finally, at the top, the *Presentation* layer includes the mechanisms to build intuitive and visual representations of the mental states of the users, abstracting from the complexity of the data level where they are positioned. This allows teachers to take efficient and intuitive decisions regarding the management of their students.

3   Case Study

In order to assess the feasibility of the tool for the intended purposes a case study was conducted in which 34 students volunteered to participate. These students participated in several tasks assigned by the teacher, of different kinds, to be solved using the computer and during one day of classes. During this day, students interacted with the computer in order to carry out their tasks, with data about their mouse and keyboard actions being recorded. Data was collected in four moments during the day in order to detect variations of the factors being studied, caused by the external factors (e.g. mental fatigue increases as the day progresses, result of our Circadian Rhythm but also of continued cognitive load and effort for attention).

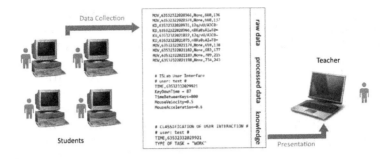

Fig. 1. Flow of data in the monitoring tool

A socio economic characterization of the population was also conducted (Figure 2) together with an analysis of possible limitations in terms of vision and hearing, which may significantly influence the students' performance. In both aspects participants evidenced no problems. Concerning physical limitations, the sample of 34 elements does not have any problems or limitations as well.

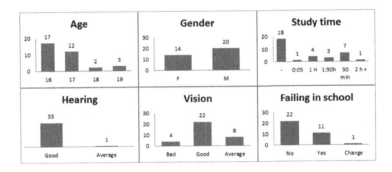

Fig. 2. Socio economical characterization of the population of the study

Another interesting variable is academic background: 65% of the students has never repeated a year. So the probability of the sample being constituted mostly by students with good grades is high. The variable "study time" reveals that the majority of students does not spend any time studying, which is however not a positive aspect. It can be seen that in general there is academic success. However this is a group of teenagers prone to several distraction factors. As external solicitations tend always too be more interesting than tasks that the student needs to achieve, this could lead to distraction, stress, and possibly fatigue. These external solicitations must then be monitored and managed.

3.1   Statistical Data Analysis

In order to verify the effect of the external factors on each feature throughout the day, data was divided in groups according to the different moments of data collection. Statistical significant tests were then conducted to assess the significance of the eventual differences to be observed between these groups of data.

Given that most of the distributions of the data are not normal, the Mann-Whitney test is used to perform the analysis. The null hypothesis is thus: $H_0 =$ the medians of the two distributions are equal. For each two distributions compared, the test returns a p-value, with a small p-value suggesting that it is unlikely that H_0 is true. Thus, for every Mann-Whitney test whose p-value $<$ α, the difference is considered to be statistically significant, i.e., H_0 is rejected. In this work, an $\alpha = 0.05$ was used.

In this specific case study, some of the features evidence statistically significant differences when comparing the different moments of the day. For each feature and for each of the moments of data collection, the average and median values were analysed in order to determine the trend of the value, i.e., we wanted to answer the question "does it tend to decrease or to increase under fatigue?"

Figure 3 depicts the distribution of the data for two features (writing velocity on the left and time between keys on the right), taken as example from the data collected during the case study. It shows how writing velocity tends to decrease and time between keys tends to increase, i.e., students write slower as activities progress throughout the day.

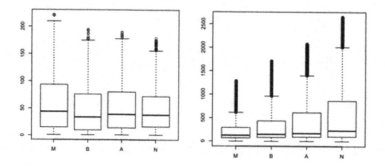

Fig. 3. Distribution of the data for the features Writing Velocity (left) and Time Between Keys (right)

An analysis of the trend of all features and all students was also conducted. Trend analyses allow to understand how students tend to behave throughout the day. The two most marked trends observed are detailed in Table 2. The mean and median values of the writing velocity tends to decrease in 93% and 89% of the students, respectively. The mean and median values of the time between keys tends to increase for 59% and 91% of the students, respectively.

These differences between the analysed moments show that as students spend more time working, external factors such as stress, fatigue or attention have a measurable effect on their interaction patterns. Essentially, the trend is towards a decrease in the performance of interaction, which was expected before hand. Indeed, as the day progresses, the prolonged effort to maintain attention to tasks

Table 2. Summary of the collected data for the two most significantly affected features

Feature	Measure	Begin	End	Trend
Writing Velocity	Mean	68.8	64.0	Decreases in 93%
	Median	46.0	41.0	Decreases in 89%
Time between keys	Mean	509.6	809.3	Increases in 59%
	Median	150	287	Increases in 91%

gradually wears the brain out. This makes it more difficult for students to focus on their task, resulting in increasing feelings of frustration, increasing stress. Moreover, a tired brain will show it through feelings of boredom or sleepiness, resulting in mental fatigue. All this affects the behaviour of students, namely in the form they interact with the computer.

While these effects were expected, the true important result in this work is to find that there is a measurable and statistically significant effect on the interaction patterns with the computer, opening the door to the development of real-time and non-invasive tools for assessing the state of students.

4   Conclusions and Future Work

In this paper we presented an ongoing work that non-intrusively monitors the effects of fatigue, stress and prolonged attention on the performance of e-learning students. Through the use of keyboard and mouse, that act as performance sensors, it is possible to estimate the influence of such factors using a specifically developed tool that monitors interaction patterns in real-time and non-intrusively. Using this information, the teacher has the possibility to assess the performance of students individually, in real time. In the future, we envision the development of recommendation systems that can provide higher-level information to the teacher, further improving his decision-making processes concerning student management and intervention. The goal is to achieve more efficient systems that analyse each student in terms of their individual characteristics, proposing and adapting specific strategies, to a specific student, in a given moment, with the aim of minimizing stress, fatigue and distractions. This work will be broadened to a much larger population, and will include mobile devices, as they are often used in learning nowadays and provide interesting feature such as accelerometer, cameras and other sensors that are prone to be used for such a context.

Acknowledgements. This work was part-funded funded by the [14VI05] Contract-Programme from the University of Vigo, by the ERDF - European Regional Development Fund through the COMPETE Programme (operational programme for competitiveness) and by Portuguese National Funds through the FCT (Portuguese Foundation for Science and Technology) within project FCOMP-01-0124-FEDER-028980 (PTDC/EEI-SII/1386/2012) and project PEst-OE/EEI/UI0752/2014.

References

1. Palmer, S., Cooper, C., Thomas, K.: Creating a Balance: Managing Stress. British Library, London (2003)
2. Rodrigues, M., Fdez-Riverola, F., Novais, P.: Moodle and Affective Computing - Knowing Who's on the Other Side. In: ECEL-2011 - 10th European Conference on Elearning, pp. 678–685 (2011)
3. Alzaghoul, A.F.: The implication of the learning theories on implementing e-learning courses. The Research Bulletin of Jordan 2(2), 27–30 (2012)
4. Hwang, K., Yang, C.: Automated Inattention and Fatigue Detection System in Distance Education for Elementary School Students. Journal of Educational Technology & Society 12, 22–35 (2009)
5. Pimenta, A., Carneiro, D., Novais, P., Neves, J.: Monitoring Mental Fatigue through the Analysis of Keyboard and Mouse Interaction Patterns. In: Pan, J.-S., Polycarpou, M.M., Woźniak, M., de Carvalho, A.C.P.L.F., Quintián, H., Corchado, E. (eds.) HAIS 2013. LNCS, vol. 8073, pp. 222–231. Springer, Heidelberg (2013)
6. Jones, F., Kinman, G.: Approaches to Studying Stress. In: Jones, F., Bright, J. (eds.) Stress: Myth, Theory and Research, Pearson Education, Harlow (2001)
7. Pimenta, A., Carneiro, D., Novais, P., Neves, J.: Analysis of Human Performance as a Measure of Mental Fatigue. In: Polycarpou, M., de Carvalho, A.C.P.L.F., Pan, J.-S., Woźniak, M., Quintian, H., Corchado, E. (eds.) HAIS 2014. LNCS, vol. 8480, pp. 389–401. Springer, Heidelberg (2014)
8. Monrose, F., Rubin, A.: Authentication via keystroke dynamics. In: Proceedings of the 4th ACM Conference on Computer and Communications Security, pp. 48–56. ACM, New York (1997)
9. Alves, F., Pagano, A., Da Silva, I.: A new window on translators' cognitive activity: methodological issues in the combined use of eye tracking, key logging and retrospective protocols. Copenhagen Studies in Language (38), 267–291 (2010)
10. Rodrigues, M., Gonçalves, S., Carneiro, D., Novais, P., Fdez-Riverola, F.: Keystrokes and Clicks: Measuring Stress on E-learning Students. In: Casillas, J., Martínez-López, F.J., Vicari, R., De la Prieta, F. (eds.) Management Intelligent Systems. AISC, vol. 220, pp. 119–126. Springer, Heidelberg (2013)
11. Turner, R.: The handbook of operator fatigue. Ergonomics 56(9), 1486 (2013)
12. Dholi, P.R., Chaudhari, K.P.: Typing Pattern Recognition Using Keystroke Dynamics. In: Das, V.V., Chaba, Y. (eds.) AIM/CCPE 2012. CCIS, vol. 296, pp. 275–280. Springer, Heidelberg (2013)
13. Kraan, K.O., Dhondt, S., Houtman, I.L.D., Batenburg, R.S., Kompier, M.A.J., Taris, T.W.: Computers and types of control in relation to work stress and learning. Behaviour & Information Technology 33(10), 1013–1026 (2014)
14. Baños, R.M., Etchemendy, E., Castilla, D., García-Palacios, A., Quero, S., Botella, C.: Positive mood induction procedures for virtual environments designed for elderly people. Interacting with Computers 24(3), 131–138 (2012)
15. Seyle, H.: The stress of life. McGraw-Hill, New York (1956)

Towards Children-Oriented Visual Representations for Temporal Relations

Tania Di Mascio and Laura Tarantino

Università degli Studi dell'Aquila, L'Aquila, Italy
{tania.dimascio,laura.tarantino}@univaq.it

Abstract. Developing the capabilities to read and comprehend text is funda-
mental for the development of children. Traditionally, the comprehension process is stimulated by educational interventions carried out by primary school
educators, who aim, e.g., at retracing temporal relations among main events of a
story. While a dual-coding approach pairing verbal and pictorial information
proves to be successful, existing proposals for the visualization of a story's
events and their relationships seems dedicated mostly to computational linguists
or information engineers rather than children and educators. The FP7 European
project TERENCE faced this issue creating the first adaptive learning system
for text comprehension for primary school children. The paper, after a review
on the state-of-art of visual representation of temporal relations, discusses the
TERENCE choices for achieving a children-oriented approach.

Keywords: Technology Enhanced Learning, temporal relations, visual interface.

1 Introduction

Reasoning coherently with time concepts (e.g., before and after) is a cognitive ability
that children have to possess in order to be proficient with text comprehension [16].
This ability starts to develop after the age of 5 and continue to develop from the age
of 7 to that of 9, when children are able to master the while connective, and futher to
the age of 11, when children become independent readers [13]. More and more children in that age range turn out to be poor (text) comprehenders, demonstrating difficulties in, among others, coherent use of connectives (because, after, before) and
inference-making from different parts of the text [6].

Text comprenhension may be improved by educational interventions aimed at reasoning about stories specifically designed so to include appropriately interspersed
temporal connectives through which children construct relations about story's events
(e.g., [5], [19]). It has to be observed that according to the dual-coding theory [18]
both verbal information and visual imagery – processed differently and separately
along distinct channels in the human mind – can be integrated to represent information so that they re-inforce each other in the learning process.

It is worth noting that the adequacy of a visual representation of time strongly depends on its final users and their tasks [1]. The issue of selecting/designing adequate

© Springer International Publishing Switzerland 2015
T. Di Mascio et al. (eds.), *Methodologies & Intelligent Systems for Technology Enhanced Learning*,
Advances in Intelligent Systems and Computing 374, DOI: 10.1007/978-3-319-19632-9_11

time representations has be faced within the context of Learning Management Systems, offering functionalities to deliver, track, report on and manage learning content, learners' progresses and learners' interactions, and of the more specific class of Adaptive Learning Systems (ALS's), able to tailor their behavior to the individual learner [4].

One such case is the TERENCE system, an ALS designed within the framework of an FP7 EU multidisciplinary project (www.terenceproject.eu), to support 7-11 years old poor text comprehenders in their learning activities and educators in the design and realization of learning material. The system presents to children adequate stories, organised into difficulty categories and collected into books, along with instructional smart games for reasoning about stories. The presentation of the learning material is actually organised as a cognitive stimulation plan designed by the neuro-psychologists involved in the project. While a comprehensive description of the "TERENCE solution" (psycho-pedagogical approach, internal and external models of stories and games, and system architecture) can be found in [11], here we focus on the requirements of children-oriented visual representrations of temporal information specifically designed for the task of story comprehension. To this aim, in Section 2 we review existing proposals for the representation of temporal connectives and assess them with reference to children-oriented design. In Section 3 we present the main features of the "read and play" visual interaction enviroment of TERENCE, and finally, in Section 4 a brief discussion on the evaluation of our proposal is given and conclusions are drawn.

2   Visual Representation of Temporal Connectives

Defining suitable visual representations for temporal structures of stories needs the integration of theoretical and methodological work both from traditional areas devoted to temporal representation (logic, reasoning, and databases) and from information visualization research field [1].

2.1   Existing Proposals for Representation of Temporal Structures

Broadly speaking, it is necessary to address issues referring to two cases: (1) single pairs of temporal events and their relations, and (2) temporal events and their relations within a whole story with more than two events.

With reference to case (1), first proposals for visualization techniques, based on seminal Allen's work on temporal logic [2], came from [14,15] (see Figure 1). Stemming from their results, other researchers proposed visual metaphors and tools surveyed e.g., in [1], [7], [9]. More specifically, [7] presented results classified according to four design dimensions: temporal structure, order, representation, and history; [9] classified literature proposals according to time granularity and capability of expressing disjunctive relations between pairs of intervals or points, while [1] uses time, data and representation dimensions.

Moving from the single pair of events of case (1) to the complexity of case (2) means moving from linear representations to graph and network representations. A proposal bridging cases (1) and (2) is in [9], who introduces three alternative visual metaphors that can scale up to the relation visualization in a network with more than two intervals. These metaphors, based on concrete objects and phenomena (elastic bands, springs and paint strips in Figure 2), can also render networks of more than two events, and their relations (case (c) in Figure 2).

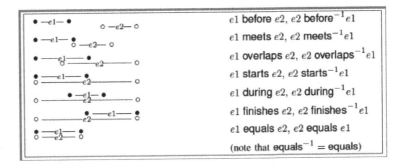

Fig. 1. Relation between intervals by [15]

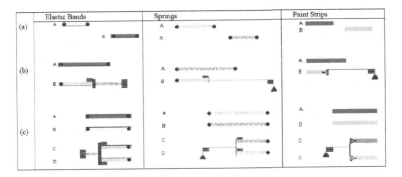

Fig. 2. The metaphorical representations proposed by [9]

As to case (2) it has to be observed that *temporal annotation* of stories is a complex task posing specific challenges deriving, among others, from high density of temporal relations and distances within the text of involved objects (e.g., events) [21]. Several tools have been proposed to deal with the complexity of temporal annotation and the consequent low markup speed, hard-to-avoid inconsistencies, and low inter-annotator agreement: in PatternFinder tool [12], temporal annotations are visualized using rows; BAT[1] describes annotations as graphs with edges and labeled nodes that are displayed according to the pivot table format; a number of proposals are based on Graphviz[2], an open source graph visualization software, among which TANGO[3] and tools derived from it, namely T-BOX [21] and TARSQI[4], and TATOT [10], which, differently from

---

[1] http://www.timeml.org/site/bat/
[2] www.graphviz.org/
[3] http://www.timeml.org/site/tango/
[4] http://www.timeml.org/site/tarsqi/toolkit/

the other tools, allows users to find a node in the graph (along with all its relations) clicking on the event annotated in the text.

2.2   Towards Children-Oriented Representations

While adequate for computational linguists or information engineers, existing visual representations fail to directly support the requirements of educators and, above all, young learners for a number of drawbacks:

- *High degree of abstraction.* Almost all proposals are based on abstract elements, and even metaphors based on real objects, such as paint strips, turns out to be not realistic enough, while recent studies prove that children appreciate realistic illustrations [8].
- *Fine granularity.* Based on Allen's work on temporal logic, all proposals aim at clearly differentiate among distinct configurations of non overlapping intervals (the first two cases in Figure 1, i.e., relations *before* and *meets*), or overlapping intervals (the other cases in Figure 1, i.e., relations *overlaps, starts, during, finishes, equals*), arriving to a level of detail that turns out to be excessive and counterproductive for children novice in the mastering of the general meaning of 'before', 'after' and 'while' [16].
- *Lack of global vision.* None of existing proposal aims at relating events to the global context of a narration, while stories are a first class tool of a psychopedagogical stimulation plan [20].
- *Groundness on a pre-existing model.* Visual representation of temporal data and temporal connectives are generally based on (crowded) timelines and visualization of intervals. Anyhow, it has to be underlined the crucial difference between adult-oriented time representation and child-oriented time representation aimed at supporting text comprehension: in the first case we build on a pre-existent mental model of time and temporal connectives, while in the latter we have to induce the construction of a mental model of time and temporal connectives. Relying on his/her mental model, an adult can grab temporal information from the exploration (and possibly the filtering) of timelines (often characterized by huge quantity of data) that would on the contrary overwhelm a child who is trying to acquire common sense time concepts. Furthermore this would be inconsistent with consolidated pedagogical approaches built on question-based games [11][20].

A final observation, more general and related to the design of any interactive application, refers to the *lack of juiciness*, and in general to the different approaches of adult and children with respect to interactions: in adult-oriented systems the focus is on productivity, with a consequent requirement of minimality of the interface in order not to distract users from their tasks, while in children-oriented application the focus is on playfulness, with a requirement of juiciness of the interface, and the achievement of the task is a side-effect of the activities carried on by the child.

For all these reasons, in TERENCE individual visualizations of events and of their temporal relations are associated to smart games and playing activities, within a visual interaction environment where children read stories and play games designed so to force the acquisition of temporal reasoning.

3   The TERENCE Proposal

The main idea behind TERENCE is that the stimulation by the system integrates with the traditional stimulation by teachers. According to the advice of the experts involved in the project, the psycho-pedagogical stimulation plan in TERENCE, based on the constructivist pedagogical approach [17], is inspired by a traditional teaching strategy including reading the story and analyzing the text via inference-making question answering. Learning sessions mirrors a warm-up, peak, and relaxing phases structure, specifically: (1) reading a story, silently – warm-up, (2) resolving related smart games for analyzing the story – peak, and (3) playing with other games able to relax the learners according to a their performances in the previous step – relaxing.

Learning material includes stories and associated smart games, along with accessory material: *stories*, organized in books, are ordered and actually written into four different versions with increased cognitive difficulty [3]; instructional *smart games* are factual (e.g., "guess who did something"), temporal (e.g., "what happened before/after this event?"), and causal (e.g., "what caused this?", or "which is the effect of this?"); *accessory material* includes elements designed in order to make the learning experience appealing, such as avatars available for the children, cards illustrating the characters of the books, relaxing games that can be played by children after the stimulation for entertainment and relaxing purposes.

Table 1. The story grammar proposed by [20]

Story Grammar based on Stein and Glenn (1979)		
Element	**Definition**	**Example**
Setting	Introduction	*Once upon a time there were three bears, the momma bear, the popa bear, and the baby bear. They all lived in a tiny house in a great big forest.*
Initiating Episode	An episode that sets up a problem or dilemma for the story	*One day a little girl named Goldilocks came by.*
Change episode	The turning episode	*She was surprised to see the house and noticed it was empty.*
Resolving episode	An action or plan of the protagonist to solve the problem	*She went inside to find the three bears gone and ate the baby bear's soup, broke the baby bear's chair, and fell asleep in the baby bear's bed.*
Final episode		*Seeing the three bears, Goldilocks ran away.*

As to story structures, the most common strategies for stimulating reading comprehension by educators are based on analysis of stories structured according to the so-called "story grammar" reported in [20] and summarized in Table 1. One may notice some key structural characteristics (SC): (SC1) Stories are relatively "light", i.e., with a limited number of episodes, events, and participating characters; (SC2) episodes are temporally ordered, and (SC3) episodes are short.

Accordingly, in TERENCE stories are structured as interactive sequences of illustrated episodes visualized according to a *carousel pattern* (Figure 3-(a)), a focus+context interactive pattern that proves to be efficient for exploring small sets (coherently with SC1). In the case of story episodes it allows the child to focus on single episodes (coherently with SC3) while maintaining a global vision on the whole story and on the order of episodes, by providing a direct visual representation of the relative time of occurrence of episodes (coherently with SC2), as illustrated in Figure 3-(b).

(a)   (b)

Fig. 3. Browsing through story episodes

As to the individual representation of events and their temporal relations, as said before, TERENCE associates them to smart games. More specifically, *time games* can be classified into:

- *before*, *after*, and *before-after* games, where the child has to reason about purely sequential events, and
- *before-while*, *while-after*, and *before-while-after*, where the child has to reason about both sequentiality and contemporaneity.

In all cases the child has to select the correct answer in a set of three. For all games the visula interface is based on the same general pattern (see examples in Figure 4). A main content area (on the right) is divided into three portions: a lower bar displays three cards corresponding to the possible choices, a middle area displays the question to be answered, and an upper part depends on the specific games.

As to the mechanisms designed so that children build their mental model of temporal relationships, we observe that, though events have a duration, and in principle might hence be associated to time intervals, the level of granularity and the degree of indeterminacy of temporal information in learner-oriented stories make interval-based visualizations (such as techniques based on Allen's relations) not adequate. In TERENCE we adopted a simple dual-coded card-based representation for the events, including an illustration and a verbal sentence. Figure 4 shows examples of a *before-after* game and a *before-while* game, based on intuitive visual metaphors for representing event sequentiality, event contemporaneity, correct answer and wrong answer. In particular, as to before/after connectives, we maintain the sequentiality of Allen's relationship while ignoring the distinction between 'before and 'meets' cases (see Figure 4(a) and Figure 4(b)); as to while connectives, we maintain Allen's suggestion to use parallelisms while ignoring the distinction among different cases of partial and complete overlapping (see Figure 4(c) and Figure 4(d)).

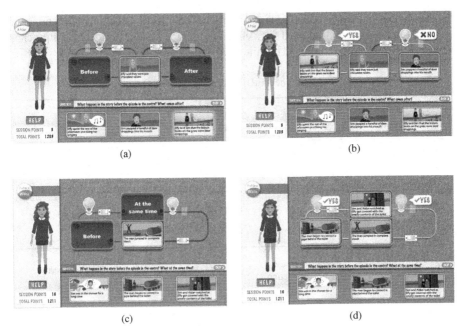

(a) (b)

(c) (d)

Fig. 4. Examples of before-after and before-while games

4   Discussion and Conclusions

In this paper we focused on one of the most important aspect for the TERENCE scientific impact: the visual representation of temporal relations. The value of this aspect was considered within the evaluations of the project. It is worth noting that TERENCE is designed and developed in an iterative manner, based on the user-centred design approach, in which four evaluations were performed: two expert-based evaluations and two user based evaluations. The complexity of the evaluation is due to the fact that, in a Technology Enhanced Learning system like TERENCE the evaluation deals with two main facets: the usability of the system and the psycho-pedagogical intervention, which investigates the learning outcomes. Table 2 provides a synoptic view of the evaluations: the first column specifies the characteristics of the evaluation (expert-based/user-based, formative/summative, qualitative/quantitative), the second and third column specify the issues and the release under evaluation, respectively, and finally the last column indicates the number of users involved.

In particular, the aspect the paper deals with is validated by considering the educational value that the TERENCE system produced; in fact, the analysis of data, gathered in a pre-post test involving two groups of users (experimental and control groups) during the summative evaluation (last column of Table 2) showed that (1) the stimulation plan significantly improved comprehension in the experimental group, (2) TERENCE improved reading comprehension also in comparison with a control group, and (3) TERENCE improved comprehension both in poor and good comprehenders, demonstrating, de facto,

that the choices implemented in the visual interface in terms of visual representations for
temporal relations are appropriated for a children-oriented system (for details on psycho-
pedagogical aspects we refer to [11], while for the main usability aspects we refer to [8]).

Table 2. The TERENCE project evaluation

Evaluation Characteristics	Issues	Release (Month)	Involved Users
1st expert-based Formative Qualitative	Usability Curricular material (e.g., stories)	Prototypes (March 2012)	about 10 domain experts of text comprehension and interaction design
1st user-based Formative Qualitative	Usability Learning outcomes	1st release (June 2012)	about 170 learners (deaf and hearing)
2nd expert-based Formative Qualitative	Smart Games revision and production	2nd release (Sept 2012)	about10 domain expert of pedagogy
2nd user-based Summative Quantitative &Quantitative	Usability Learning outcomes	3rd release (March 2013)	About 830 learners (deaf and hearing)

References

1. Aigner, W., Miksch, S., Muller, W., Schumann, H., Tominski, C.: Visualizing Time-Oriented Data, A Systematic View. J. Computers and Graphics 31, 401–409 (2007)
2. Allen, J.F.: Maintaining Knowledge about Temporal Intervals. ACM Comm. 26, 832–843 (1983)
3. Arfé, B., Oakhill, J., Pianta, E.: The Text Simplification in TERENCE. In: Di Mascio, T., Gennari, R., Vitorini, P., Vicari, R., de la Prieta, F. (eds.) Methodologies and Intelligent Systems for Technology Enhanced Learning. AISC, vol. 292, pp. 165–172. Springer, Heidelberg (2015)
4. Brusilovsky, P., Millán, E.: User Models for Adaptive Hypermedia and Adaptive Educational Systems. In: Brusilovsky, P., Kobsa, A., Nejdl, W. (eds.) Adaptive Web 2007. LNCS, vol. 4321, pp. 3–53. Springer, Heidelberg (2007)
5. Cain, K., Nash, H.M.: The Influence of Connectives on Young Readers' Processing and Comprehension of Text. J. of Educational Psychology 103, 429–441 (2011)
6. Cain, K., Oakhill, J.V., Elbro, C.: The Ability to Learn New Word Meanings from Context by School-Age Children with and Without Language Comprehension Difficulties. J. of Child Language 30, 681–694 (2003)
7. Catarci, T., Fernandes Silva, S.: Visualization of Linear Time-Oriented Data: a Survey. In: Proceedings of the First International Conference on Web Information Systems Engineering, pp. 310–319. ACM Press (2000)

8. Cecilia, M.R., Di Mascio, T., Tarantino L., Vittorini, P.: Designing TEL products for poor comprehenders: evidences from the evaluation of TERENCE. Interaction Design and Architecture, Special Issue on The Design of TEL with Evidence and Users (in press)

9. Chittaro, L., Combi, C.: Representation of Temporal Intervals and Relations: Information Visualization Aspects and their Evaluation. In: Proceedings of the 8th International Symposium on Temporal Representation and Reasoning, pp. 13–20. IEEE Press, Los Alamitos (2001)

10. Di Mascio, T., De Gasperis, G., Florio, N.: TATOT: a viewer for annotate stories. In: Proceedings of ITAIS (2011), http://www.cersi.it/itais2011/pdf/72.pdf

11. Di Mascio, T., Gennari, R., Melonio, A., Tarantino, L.: Supporting children in mastering temporal relations of stories: the TERENCE learning approach. Intl. J. on Distance Education Technology, Special Issue on The Design of TEL with Evidence and Users (in press)

12. Fails, J.A., Karlson, A., Shahamat, L., Shneiderman, B.: A Visual Interface for Multivariate Temporal Data: Finding Patterns of Events across Multiple Histories. In: Wong, P.C., Keim, D.A. (eds.) Visual Analytics Science and Technology, pp. 167–174. IEEE Press (2006)

13. Ge, F., Xuehong, T.: Temporal Reasoning on Daily Events in Primary School Pupils. Acta Psychological Sinica 34, 604–610 (2002)

14. Hajnicz, E.: Time Structures: Formal Description and Algorithmic Representation. LNCS (LNAI), vol. 1047. Springer, Heidelberg (1996)

15. Hibino, S., Rundensteiner, E.A.: User Interface Evaluation of a Direct Manipulation Temporal Visual Query Language. In: Proceedings of the ACM Conference on Multimedia, pp. 99–107. ACM Press (1997)

16. McColgan, K., McCormack, T.: Searching and Planning: Young Children's Reasoning about Past and Future Event Sequences. Child Development 2, 1477–1497 (2008)

17. Nanjappa, A., Grant, M.M.: Constructing on constructivism: The role of technology. Electronic Journal for the Integration of Technology in Education 2, 38–56 (2003)

18. Paivio, A.: Dual-coding Theory: Retrospect and Current Status. Canadian Journal of Psychology 45, 255–287 (1991)

19. Trabasso, T., Van den Broek, T.: Causal Thinking and the Representation of Narrative Events. J. of Memory and Language 24, 612–630 (1985)

20. Valeriani, A.: Ermeneutica retorica ed estetica nell'insegnamento verso l'oriente del testo. Andromeda, Teramo (1986)

21. Verhagen, M.: Drawing TimeML Relations with TBox. In: Schilder, F., Katz, G., Pustejovsky, J. (eds.) Annotating, Extracting and Reasoning about Time and Events. LNCS (LNAI), vol. 4795, pp. 7–28. Springer, Heidelberg (2007)

Author Index

Printed in the United States
By Bookmasters